Avalonian and Cadomian Geology
of the North Atlantic

Avalonian and Cadomian Geology
of the North Atlantic

Edited by

R. A. STRACHAN and G. K. TAYLOR

Department of Geology
Oxford Polytechnic

Blackie
Glasgow and London

Published in the USA by
Chapman and Hall
New York

Blackie & Son Limited.
Bishopbriggs, Glasgow G64 2NZ
and
7 Leicester Place, London WC2H 7BP

Published in the USA by
Chapman and Hall
a division of Routledge, Chapman and Hall, Inc.
29 West 35th Street, New York, NY 10001-2291

British Library Cataloguing in Publication Data

Avalonian and Cadomian geology of the North Atlantic.
1. North Atlantic region. Strata
I. Strachan, R. A.   II. Taylor, G. K.
551.70091631

ISBN 0-216-92686-6

Library of Congress Cataloging-in-Publication Data

Avalonian and Cadomian Geology of the North Atlantic   /   edited by
R. A. Strachan, and G. K. Taylor.
     p.   cm.
  Bibliography: p.
  Includes index.
  ISBN 0-412-02151-X (Chapman and Hall)
  1. Geology, Stratigraphic—Precambrian. 2. Geology—North
Atlantic Ocean Region.   I. Strachan, Richard A.   II. Taylor, G. K.
QE653.P36   1990   89-15828
                                                              CIP

Filmset by Advanced Filmsetters (Glasgow) Ltd
Printed in Great Britain by Thomson Litho Ltd, East Kilbride, Scotland

Preface

Avalonian and Cadomian Geology of the North Atlantic aims to provide region by region syntheses of those terranes which are believed to have once formed part of the Late Proterozoic Avalonian–Cadomian belt. The belt originally comprised a series of volcanic arcs and marginal basins formed during the period *c*. 700–500 Ma ago, producing a range of calc-alkaline volcanics and intrusives and marginal basin sediments partly founded on continental basement rocks. Originally about 9000 km long, the belt is made up of a number of terranes later separated by the Caledonian orogeny (*c*. 500–400 Ma), Variscan orogeny (*c*. 325 Ma), and the recent opening of the Atlantic Ocean (*c*. 70 Ma). The rocks therefore are now located on both sides of the Atlantic, as far apart as Florida and Czechoslovakia. A possible modern analogue for the Avalonian–Cadomian belt may be the present-day Western Pacific margins. We believe that a synthesis such as this provides the basis for correlation of the timing of events and hence improves our understanding of the underlying tectonic framework of the belt.

This book developed from the recognition of, first, our own lack of knowledge of contemporaneous events outwith Armorica, and, second, the likelihood that other researchers working on individual areas or terranes do so in relative isolation. We have therefore commissioned chapters from researchers considered to be experts on their own particular areas. Although many problems remain to be resolved, we have attempted a synthesis of the data and have presented a simplistic tectonic model for the evolution of the belt. There remains much scope for the improvement and refinement of our model, but it should stimulate further discussion and research, and help with the formulation of better models in the future.

Given the diversity of expertise of individual contributors, each chapter contains a blend of results reflecting the wide and varying range of techniques currently applied to terrane analysis, e.g. structural studies, sedimentology, geochemistry, geophysics and geochronology. Also, given the diversity, complexity, and varying levels of knowledge of individual terranes, we encouraged contributors to emphasise the most important events which are recorded in each area, and to provide comprehensive and up-to-date references for those readers who wish to explore the finer points in more detail. We hope that this volume provides a stimulus and foundation for future research which will ultimately improve our understanding of the geological evolution of the Avalonian–Cadomian belt.

RAS
GKT

Contributors

J. Chaloupský Geological Survey, Prague, Malostranske nam. 19, Prahal, Czechoslovakia.

K. L. Currie Geological Survey of Canada, 601 Booth Street, Ottawa, Ontario K1A 0E8, Canada.

R. D. Dallmeyer Department of Geology, University of Georgia, Athens, Georgia 30602, USA.

J. Dostal Department of Geology, St. Mary's University, Halifax, Nova Scotia B3H 3C3, Canada.

W. Gibbons Department of Geology, University of Wales, Cardiff CF1 3YE, UK.

J. D. Keppie Department of Mines and Energy, PO Box 1087, Halifax, Nova Scotia B3J 2X1, Canada.

A. F. King Department of Earth Sciences, Memorial University of Newfoundland, St. John's, Newfoundland A1B 3X5, Canada.

M. D. Max Naval Research Laboratory, Washington DC, 20375-5000, USA.

F. C. Murphy Department of Geology, University College, Dublin 4, Ireland.

J. B. Murphy Department of Geology, St. Francis Xavier University, Antigonish, Nova Scotia B2G 1C0, Canada.

R. D. Nance Department of Geological Sciences, Ohio University, Athens, Ohio 45701, USA.

S. J. O'Brien Department of Mines and Energy, Geological Survey Branch, PO Box 8700, St. John's, Newfoundland A1B 4J6, Canada.

J. C. Pauley Poroperm-Geochem Limited, Chester Street, Chester CH4 8RD, UK.

C. Quesada   ITGE, Rios Rosas 23, 28003 Madrid, Spain.

R. A. Roach   Department of Geology, University of Keele, Keele
ST5 5BG, UK.

R. A. Strachan   Department of Geology, Oxford Polytechnic, Oxford
OX3 0BP, UK.

D. F. Strong   Department of Earth Sciences, Memorial University of
Newfoundland, St. John's, Newfoundland A1B 3X5, Canada.

G. K. Taylor   Department of Geology, Oxford Polytechnic, Oxford
OX3 0BP, UK.

P. J. Treloar   Department of Geology, Imperial College, London
SW7 2BP, UK.

J. A. Winchester   Department of Geology, University of Keele, Keele
ST5 5BG, UK.

Contents

1 **Introduction**   **1**
R. A. STRACHAN and G. K. TAYLOR

    References   3

2 **The Longmyndian Supergroup and related Precambrian sediments of England and Wales**   **5**
J. C. PAULEY

    2.1   Introduction   5
    2.2   Longmyndian Supergroup, Welsh Borderland   5
       2.2.1   Stratigraphy of the Longmyndian Supergroup   7
       2.2.2   Structure of the Longmyndian Supergroup   9
       2.2.3   Sedimentology of the Longmyndian Supergroup   10
    2.3   Old Radnor Inlier, Welsh Borderland   12
    2.4   Charnian Supergroup, Leicestershire   15
    2.5   Llangynog Inlier, Dyfed   18
    2.6   Other outcrops of Precambrian sediments in England and the Welsh Borderland   19
    2.7   Depositional setting of the Precambrian sediments   21
    References   24

3 **Pre-Arenig terranes of northwest Wales**   **28**
W. GIBBONS

    3.1   Introduction and historical perspective   28
    3.2   The Monian Supergroup   33
    3.3   The Coedana Complex   36
    3.4   The southeast Anglesey blueschists   38
    3.5   The Sarn Complex   40
    3.6   Regional correlations   40
    3.7   Overstep sequences and age constraints   41
    3.8   Monian tectonics: an overview   44
    References   46

4 **The Rosslare Complex: a displaced terrane in southeast Ireland**   **49**
J. A. WINCHESTER, M. D. MAX and F. C. MURPHY

    4.1   Introduction   49
    4.2   The Rosslare Complex: constituents and geological history   51
       4.2.1   Early tectono-thermal history   51
       4.4.2   The second event (M2)   55
       4.2.3   The third event (M3)   56
       4.2.4   The fourth event (M4)   57
    4.3   Rocks marginal to the Rosslare Complex   57
       4.3.1   Tomhaggard Zone   57

4.3.2   Silverspring Zone   57
4.3.3   Tagoat Group   60
4.3.4   The Ballycogly Group   60
4.3.5   Cullenstown Formation   60
4.3.6   Tuskar Group   60
4.4   Caledonian intrusions   61
4.4.1   Saltees Granite   61
4.4.2   Younger basic dykes   61
4.4.3   Carnsore Granite   62
4.5   Conclusions   62
References   63

5   Cadomian terranes in the North Armorican Massif, France   65
R. A. STRACHAN, R. A. ROACH and P. J. TRELOAR

5.1   Introduction   65
5.2   St. Brieuc terrane (SBT)   67
5.2.1   Pre-Cadomian Icartian basement   69
5.2.2   The Penthièvre complex   69
5.2.3   Brioverian supracrustals   72
5.2.4   Cadomian magmatism   74
5.3   St. Malo terrane (SMT)   76
5.4   Mancellian terrane (MT)   77
5.5   Nature and timing of deformation   78
5.6   Post-Cadomian sedimentation   81
5.7   Regional correlations   83
5.7.1   Guernsey, Sark, Alderney and Cap de la Hague   83
5.7.2   Jersey   85
5.7.3   Lower Normandy   85
5.8   Geotectonic setting   86
References   88

6   Precambrian of the Bohemian Massif, Central Europe   93
J. CHALOUPSKÝ

6.1   Introduction   93
6.2   The Bohemian Massif   93
6.3   The Moldanubian   97
6.4   The Brioverian   99
6.4.1   The Lower Brioverian   99
6.4.2   The Middle Brioverian   100
6.4.3   The Upper Brioverian   101
6.5   Unit boundaries   102
6.6   Cadomian deformation   103
6.7   Cadomian metamorphism and plutonism   103
References   106

7   Precambrian terranes in the Iberian Variscan Foldbelt   109
C. QUESADA

7.1   Introduction   109
7.2   Precambrian sequences and history of northwest Iberia suspect terranes 111
7.3   Stratigraphy of Precambrian sequences in the Iberian terrane   113
7.3.1   Pre-orogenic sequences   114
7.3.2   Syn-orogenic sequences   119

7.4 Late Precambrian tectono-thermal evolution of the Iberian terrane: the
Cadomian orogeny 122
    7.4.1 Cadomian deformational events 122
    7.4.2 Cadomian metamorphism 124
    7.4.3 Cadomian igneous activity 126
    7.4.4 The Cadomian orogeny in the Iberian terrane from a terrane
        perspective 127
References 130

8 The West African orogens and Circum-Atlantic correlatives   134

R. D. DALLMEYER

8.1 Introduction 134
8.2 Mauritanide, Bassaride, and Rokelide orogens 134
    8.2.1 Geologic setting 136
    8.2.2 Tectono-thermal history 142
    8.2.3 Geophysical controls on tectonic models 144
    8.2.4 Geodynamic models 145
8.3 Southern Appalachian orogen 147
8.4 Pre-Cretaceous crystalline basement beneath the Atlantic and Gulf
Coastal Plains of the southeastern United States 150
    8.4.1 Suwannee terrane 150
    8.4.2 Structure 153
    8.4.3 Relationship to Appalachian elements 154
    8.4.4 Wiggins Uplift 154
    8.4.5 Southwestern Alabama Igneous Complex 155
    8.4.6 Regional tectonic relations 155
8.5 Appalachian–West African correlations 156
8.6 Terrane accretion in the southern Appalachian orogen 158
References 160

9 The Avalon Zone type area: southeastern Newfoundland Appalachians   166

S. J. O'BRIEN, D. F. STRONG and A. F. KING

9.1 Introduction 166
9.2 Characteristic lithology and age 168
9.3 Boundaries 170
9.4 Precambrian stratigraphy 171
    9.4.1 Pre-635 Ma rocks 171
    9.4.2 Late Precambrian (635 Ma–600 Ma) volcanic rocks 172
    9.4.3 Late Precambrian (635 Ma–c. 570 Ma) marine sedimentary rocks 174
    9.4.4 Latest Precambrian terrestrial sedimentary rocks 176
    9.4.5 Latest Precambrian (c. 570 Ma) volcanic rocks 177
9.5 Late Precambrian plutonic rocks 178
9.6 Cambro-Ordovician stratigraphy 180
9.7 Mid-Palaeozoic stratigraphy 181
9.8 Palaeozoic plutonic rocks 182
9.9 Mesozoic magmatism 182
9.10 Tectono-thermal history 182
9.11 Synthesis 186
    9.11.1 Deep crust 186
    9.11.2 The Burin Group 186
    9.11.3 Volcanic and plutonic rocks 187
    9.11.4 Sedimentary rocks 187
    9.11.5 Latest Precambrian rocks 188

9.11.6　The Eocambrian–Palaeozoic record　188
9.11.7　The Avalonian–Appalachian connection　189
References　190

10　The Avalon Composite terrane of Nova Scotia　195
J. B. MURPHY, J. D. KEPPIE, R. D. NANCE
and J. DOSTAL

10.1　Introduction　195
10.2　Geological data　200
　10.2.1　Mid/Late Proterozoic gneisses　200
　10.2.2　Mid/Late Proterozoic platformal rocks　203
　10.2.3　Late Proterozoic volcano-sedimentary rocks　204
　10.2.4　Early Palaeozoic overstep sequences　208
10.3　Discussion　210
References　212

11　The Avalon Zone of New Brunswick　214
R. D. NANCE, K. L. CURRIE and J. B. MURPHY

11.1　Zone definition　214
11.2　Zone boundaries　214
11.3　Unit descriptions　219
　11.3.1　Brookville Gneiss　220
　11.3.2　Green Head Group　221
　11.3.3　Martinon Formation　222
　11.3.4　Early amphibolites?　224
　11.3.5　Late Precambrian granitoid plutons (Golden Grove suite)　224
　11.3.6　Coldbrook Group　225
　11.3.7　Kingston Dyke Complex(?)　227
　11.3.8　Late Precambrian (?) mylonite zones　228
　11.3.9　'Eocambrian' succession　230
　11.3.10　Saint John Group　231
11.4　Regional correlations　232
References　234

**12　Palaeomagnetic and tectonic constraints on the development
of Avalonian–Cadomian terranes in the North Atlantic region　237**
G. K. TAYLOR and R. A. STRACHAN

12.1　Introduction　237
12.2　Palaeomagnetic constraints on a late Precambrian plate reconstruction　237
　12.2.1　The main cratons　237
　12.2.2　The intervening terranes　240
　12.2.3　Comparison with previous reconstructions　241
12.3　Tectonic synthesis　242
References　246

Index
　　249

1 Introduction

R. A. STRACHAN and G. K. TAYLOR

Reconstructions of the pre-Mesozoic positions of the continents in the North Atlantic region indicate former continuity of two major Upper Proterozoic to Lower Palaeozoic orogenic belts (Figure 1.1), namely the Caledonide–Appalachian and Avalonian–Cadomian–Pan African belts which currently occupy the margins of the present-day Atlantic ocean. These belts were juxtaposed during the Palaeozoic as a result of the closure of the Iapetus and Theic oceans which led to the assembly of the supercontinent Pangea (Rodgers, 1988).

The first of these orogenic belts comprises the Appalachian–Caledonide belt of eastern North America, the northern British Isles, East Greenland, Svalbard and western Scandinavia (Figure 1.1). The autochthonous parts of this belt consists of a basement of mainly Mid- to Upper Proterozoic rift and shelf clastics and lower Palaeozoic shelf sequences which formed the eastern margin of the Laurentian–Greenland craton and the western and southern margins of the Fennoscandian craton (Figure 1.1; Skehan, 1988). During the Upper Cambrian to Lower Ordovician, volcanic arcs collided with the cratons and their sedimentary cover leading to deformation, metamorphism and ophiolite obduction. Renewed deformation during the Upper Ordovician through to the Devonian was associated with further stages of oceanic closure. Full reviews of the development of the Appalachian–Caledonide belt are provided in Harris *et al.* (1979), Gee and Sturt (1985) and Harris and Fettes (1988).

The second of these belts is the subject of this volume and comprises the Avalonian–Cadomian–Pan African belt of northwest Africa, eastern North America including the Maritime Provinces of eastern Canada and Newfoundland, Spain, northwest France, Czechoslovakia and southern Britain (Figure 1.1). This belt, except in northwest Africa, shows limited evidence of basement, but includes substantial thicknesses of variably deformed and metamorphosed Upper Proterozoic (700–500 Ma) volcano-sedimentary and plutonic complexes which are overlain unconformably by lower Palaeozoic sequences containing Acado-Baltic faunas (Skehan, 1988, and references therein). Although variable Acadian (Upper Silurian–Lower Devonian) and/or Variscan (Upper Devonian–Mid Carboniferous) over-printing has in some areas (e.g. northwest France and Spain) made

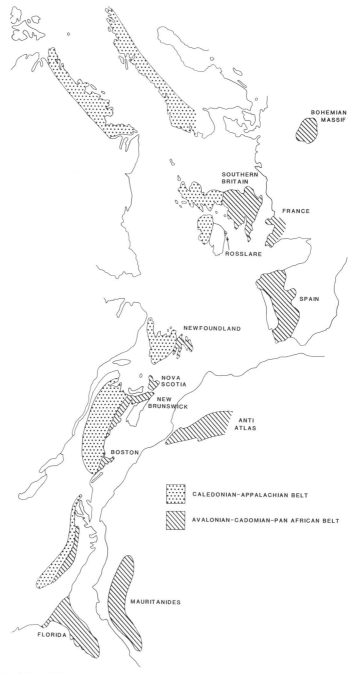

Figure 1.1 Map of the North Atlantic borderlands in their early Mesozoic pre-drift positions, showing the approximate extent of the Avalonian–Cadomian terranes and the Caledonian–Appalachian belt (modified from Skehan, 1988).

elucidation of this history difficult, it is clear that the early history of this belt is distinct in many respects from that of the Appalachian–Caledonide belt to the north and west. We believe that all the component parts of this belt may be generally considered to have been either linked or in close proximity to the Gondwanan craton during the late Precambrian/earliest Cambrian period under discussion in this volume. Constituent parts of this orogenic belt have variously been referred to as Avalonian (e.g. Newfoundland, southern Britain etc.), Cadomian (e.g. northwest France, Spain) or Pan African (West Africa).

Autochthonous parts of the Pan African belt are present in West Africa and Morocco (Figure 1.1) where c. 2000 Ma-old Eburnean basement gneisses are overlain unconformably by shelf sediments which were deformed and metamorphosed during the Pan African orogeny 680–580 Ma (Leblanc, 1981; Lecorche et al., 1983; Dallmeyer, 1989). The general opinion of most workers would seem to be that other Avalonian–Cadomian terranes acted as independent microplates during the late Precambrian and may never have been attached to Gondwana, or alternatively, they may have become detached during later Acadian or Variscan events. The presence of numerous strike-slip shear zones in southern Britain, northwest France, Newfoundland and Nova Scotia–New Brunswick implies that these terranes underwent substantial lateral displacement along the northwest margin of Gondwana during Avalonian–Cadomian–Pan African events and therefore have the status of suspect terranes relative to the autochthonous craton. Later movements along Variscan strike-slip shear zones (e.g. the North and South Armorican shear zones) may have further increased the displacement of several of these terranes.

The purpose of this volume is to review the geology of some of the most important Avalonian–Cadomian terranes in the North Atlantic region. Each chapter is written by research workers currently active in these individual terranes. The chapters are arranged geographically, commencing in southern Britain and proceeding generally southward to the Mauritanides of northwest Africa, and thence northeastwards to the Maritime Provinces of Canada (Figure 1.1). The final chapter discusses the constraints imposed on models by the available palaeomagnetic data and synthesises this information into a new tectonic model for the evolution of the Avalonian–Cadomian terranes.

References

Dallmeyer, R. D. (1989) The West African orogens and Circum-Atlantic correlatives. In *Avalonian and Cadomian Geology of the North Atlantic*, ed. Strachan, R. A. and Taylor, G. K., Chapter 8 (this volume).

Gee, D. G. and Sturt, B. A. (1985) *The Caledonide Orogen—Scandinavia and Related Areas*. Wiley, Chichester.

Harris, A. L. and Fettes, D. J. (1988) The Caledonian–Appalachian Orogen. *Spec. Publ. Geol. Soc. London* **38**.

Leblanc, M. (1981) The late Proterozoic ophiolites of Bou Azzer (Morocco): evidence for Pan-

African plate tectonics. In *Precambrian Plate Tectonics*, ed. Kroner, A., Elsevier, Amsterdam, 435–451.

Lecorche, J. P., Roussel, J., Sougy, J. and Guetat, Z. (1983) An interpretation of the geology of the Mauritanides orogenic belt (West Africa) in the light of geophysical data. In *Contributions to the Tectonics and Geophysics of Mountain Chains*, eds. Hatcher, R. D., Williams, H. and Zietz, I., *Geol. Soc. Am., Mem.* **158**, 131–147.

Rodgers, J. (1988) Fourth time-slice: Mid-Devonian to Permian Synthesis. In *The Caledonian–Appalachian Orogen*, eds. Harris, A. L. and Fettes, D. J., *Spec. Publ. Geol. Soc. London* **38**, 621–626.

Skehan, J. W. (1988) Evolution of the Iapetus Ocean and its borders in Pre-Arenig times: a synthesis. In *The Caledonian–Appalachian Orogen*, Harris, A. L. and Fettes, D. J., *Spec. Publ. Geol. Soc. London* **38**, 185–229.

2 The Longmyndian Supergroup and related Precambrian sediments of England and Wales

J. C. PAULEY

2.1 Introduction

Two thick sequences of late Precambrian sediments crop out in the Welsh Borders and central England; the Longmyndian Supergroup and the Charnian Supergroup, respectively (Figure 2.1). Small inliers of related late Precambrian sediments crop out at Old Radnor, Powys, and near Llangynog, Dyfed. It is the purpose of the following account to review these sedimentary sequences, to compare and contrast them, and to discuss their depositional setting. Of importance in the latter respect is the relationship between these sedimentary sequences and the associated Precambrian igneous and metamorphic rocks. Such igneous rocks include the Malvernian plutonic complex, the Uriconian Volcanic Complex, the Stanner-Hanter Complex, the Bardon Hill and Whitwick Complexes, the Northern and Southern Diorites, the Llangynog igneous rocks, and the Caldecote Volcanics (Figure 2.1). Metamorphic rocks crop out at Rushton, near Wellington, at nearby Primrose (Little) Hill, and in the Malvern Hills. The outcrops of Precambrian rocks usually occur along major lineaments, the most important of these being the Church Stretton and Pontesford-Linley fault systems, and the Malvern Lineament. These have exerted a great influence on the structure of the Precambrian rocks.

2.2 Longmyndian Supergroup, Welsh Borderland

The Longmyndian Supergroup crops out in the vicinity of Church Stretton, Shropshire, mainly within an area 9 km wide and 25 km long elongated NE–SW (Figure 2.2). It is bounded on its northwestern margin by the NNE–SSW trending Pontesford-Linley fault (Woodcock, 1984b), and on its southeastern margin by the NE–SW trending Church Stretton fault system. These are components of the Welsh Borderland Fault System (Woodcock and Gibbons, 1988). Precambrian igneous rocks of the Uriconian Volcanic Complex crop out within the fault systems (Boulton, 1904; Pocock et al., 1938; James, 1952a, b, 1956; Greig et al., 1968; Thorpe, 1972, 1974, 1979, 1982; Thorpe et al., 1984; Pauley, 1986, 1990; Pharaoh et al., 1987b). Small areas of

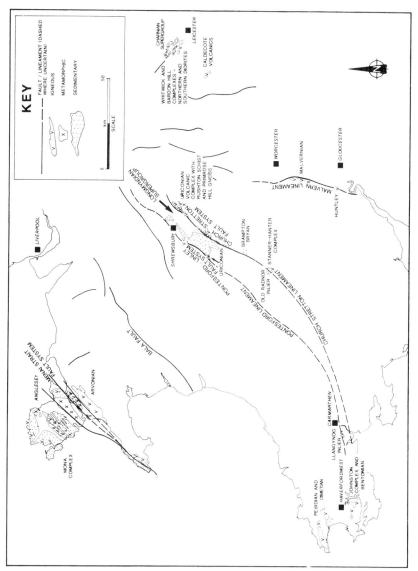

Figure 2.1   Geological map showing the Precambrian outcrops of England and Wales.

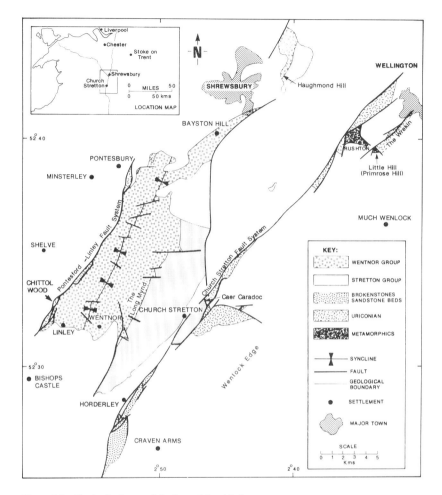

Figure 2.2 Geological map of the Long Mynd Inlier.

Precambrian metamorphic rocks crop out within the Church Stretton fault system in the vicinity of Rushton and to the southwest of the Wrekin at Primrose (Little) Hill (Pocock *et al.*, 1938; Greig *et al.*, 1968; Thorpe, 1974). The Longmyndian Supergroup comprises about 6500 m of sandstones and mudstones, with occasional conglomerates and tuffs (James, 1952a, b, 1956; Greig *et al.*, 1968; Pauley, 1986, 1990).

2.2.1 *Stratigraphy of the Longmyndian Supergroup*

The stratigraphy established by Greig *et al.* (1968) and James (1952b, 1956) has been modified by Pauley (1986, 1990). The terms Group and Series as used by Greig *et al.* (1968) have been replaced by Formation and Group, respectively, and the Longmyndian is assigned the status of Supergroup. The contentious stratigraphical issues are the relationships between the Longmyndian Supergroup and the Uriconian Volcanic Complex, and the presence or absence of unconformities within the Longmyndian Supergroup.

An unconformity between the Wentnor Group and the Stretton Group has been generally accepted by previous authors though some early authors, notably Lapworth and Watts (1910) and Callaway (1891), did not recognise an unconformity. Pauley (1986, 1990) argued that an unconformity is lacking in the main exposure of the Long Mynd and that an angular discordance between the two Groups is the result of syn- or post-folding deformation. The evidence for an unconformity between the Stretton and Wentnor Groups exposed within the Church Stretton fault system (Dean, 1964) was argued to be equivocal. Even if an unconformity were present in this case, it does not require an unconformity in the main inlier. James (1952a, b, 1956) proposed that there is an unconformity between the Wentnor Group and the Western Uriconian in the vicinity of Chittol (Figure 2.2), this being used as one argument for an unconformity at the base of the Wentnor Group. However, Pauley (1986, 1990) interpreted the contacts between the Western Uriconian and the Longmyndian Supergroup as faults which are components of the Pontesford-Linley fault system.

The Helmeth Grit has previously been considered to be the base of the Longmyndian Supergroup and to overlie the Ragleth Tuffs of the Uriconian Volcanic Complex unconformably (Cobbold and Whittard, 1935; James, 1952b, 1956; Greig *et al.*, 1968). However, Pauley (1986, 1990) argued that the Ragleth Tuffs should be incorporated within the Longmyndian Supergroup since they are of a similar sedimentary character and are distinct from the igneous rocks of the Uriconian Volcanic Complex. It was also argued that the Helmeth Grits should be included as a member of the Ragleth Tuff Formation based on their similarity which was also noted by Cobbold and Whittard (1935), and an unconformable relationship between them was disputed. Pauley (1986, 1990) argued that the Ragleth Tuff Formation is probably faulted against the Stretton Shale Formation and is therefore of uncertain stratigraphic position with respect to the Longmyndian Super- group. In addition both the Ragleth Tuff Formation and the Stretton Shale Formation were thought to be faulted against the Uriconian Volcanic Complex, the faults being components of the Church Stretton fault system. The Longmyndian Supergroup is therefore considered to be only in fault contact with the Uriconian Volcanic Complex and their stratigraphic relationship is open to interpretation.

The Longmyndian Supergroup is mainly composed of volcanic detritus and the lithologies of the clasts can be matched with some of those of the Uriconian Volcanic Complex. Therefore either the Uriconian or a similar volcanic complex acted as a source. This has been accepted by most previous authors. Pyroclastic deposits within the Stretton Group suggest that the volcanic source was active during deposition. Therefore it is thought that the Longmyndian Supergroup is partly coeval with the Uriconian Volcanic Complex.

There are isolated outcrops of conglomerate to the east of the Church Stretton Fault. This conglomerate was referred to as the Willstone Hill Conglomerate by Pauley (1986). Greig *et al.* (1968) included this with the Eastern Uriconian. However Pauley (1986) thought that it is possibly an equivalent of the Wentnor Group which overlies the Uriconian Volcanic Complex unconformably, as suggested by its textural and petrographic characteristics. If this interpretation is correct, it suggests that the Uriconian Volcanic Complex is partly older than the Longmyndian Supergroup. The Longmyndian facies and palaeocurrent data do not indicate derivation from the Uriconian Volcanic Complex in its present geographic position (Pauley, 1986, 1990). The recognition of strike-slip faulting along the Church Stretton and Pontesford-Linley fault systems (Woodcock, 1984a, b, 1988; Pauley, 1986, 1990; Lynas, 1988; Woodcock and Gibbons, 1988) suggests that this may be the result of strike-slip displacements of the Longmyndian Supergroup with respect to the Uriconian Volcanic Complex.

2.2.2 *Structure of the Longmyndian Supergroup*

The Longmyndian Supergroup has been folded into a tight, NNE–SSW trending syncline which plunges gently towards the south-southwest (James, 1952b, 1956; Greig *et al.*, 1968). Way-up evidence, based on sedimentological criteria and the relationship between cleavage and bedding, was collected by the present author (Pauley, 1986) and confirms the existence of this syncline. There are common parasitic folds at significant changes in lithology which are of a similar orientation to the main syncline. The extent of these was not previously realised and they are more common in the Stretton Group than was previously thought. There is evidence which indicates that the Longmyndian Supergroup was deformed during sinistral transpression (Pauley, 1986, 1990). The faults which cut the Longmyndian syncline are interpreted as a Riedel shear response to sinistral strike-slip parallel to the fold axis and to the Church Stretton and Pontesford-Linley fault systems. Ductile strain markers indicate a component of extension subparallel to the fold axis, which is interpreted as the result of either compression followed by transcurrent movements, or transpression.

The Church Stretton and Pontesford-Linley fault systems are interpreted as strike-slip fault systems (Woodcock, 1984a, b, 1988; Pauley, 1986, 1990; Lynas, 1988; Woodcock and Gibbons, 1988; Woodcock *et al.*, 1988). The history of strike-slip movements along these is complex. Late Ordovician to early Silurian, probably dextral, movements were recognised by Lynas (1988), Woodcock (1984a, b), and Woodcock and Gibbons (1988). Sinistral movements of Devonian (Acadian) age were proposed by Woodcock (1988), Woodcock and Gibbons (1988) and Woodcock *et al.* (1988). Post-Carboniferous and possibly post-Triassic movements were suggested by Woodcock (1984a, b). Woodcock and Gibbons (1988) suggested that the Welsh Borderland Fault System could have originated during late Precambrian to early Cambrian transcurrent movements or possibly during Cambro-Ordovician extension.

The Longmyndian syncline is likely to have originated at about the same time as the Church Stretton and Pontesford-Linley fault systems were operating in strike-slip, or shortly before this. The stratigraphic evidence only constrains this syncline to be of pre-Upper Llandovery and possibly of pre-Caradoc age. There is some stratigraphic evidence which suggests that the Church Stretton fault system originated prior to the Caradocian (Pauley, 1990): the unconformities between the late Tommotian Wrekin Quartzite and the underlying Uriconian Volcanic Complex (e.g., Cope and Gibbons, 1987) and between the Wrekin Quartzite and the Rushton Schist (Pocock *et al.*, 1938) suggest that the Church Stretton fault system originated in late Precambrian to earliest Cambrian time. A number of radiometric age-dates between *c.* 535 Ma and *c.* 520 Ma obtained from the Precambrian rocks of Shropshire by fission-track and Rb-Sr age-dating methods (Bath, 1974; Patchett *et al.*, 1980; Naeser *et al.*, 1982) point towards a possible tectonic event during this period. This is thought to have involved sinistral transpression of the Longmyndian Supergroup and sinistral strike-slip along the fault systems (Pauley, 1990). However, if this tectonic event were pre–late Tommotian in age then the radiometric age-dates appear to be too young (Cope and Gibbons, 1987) as discussed in a following section. The hypothesis of sinistral transpression during the late Precambrian to earliest Cambrian (Pauley, 1986, 1990) is supported by evidence from other areas. Dallmeyer and Gibbons (1987) and Gibbons (1987) recognised a similar deformation of the Precambrian of northwest Wales at a similar time. Murphy (1990) interpreted the mylonites of the Precambrian Rosslare Complex of southeast Ireland to be the result of sinistral transcurrent movements of probable late Precambrian age.

2.2.3 *Sedimentology of the Longmyndian Supergroup*

The sedimentology of the Longmyndian Supergroup has been discussed only briefly by previous authors (Taylor, 1958; Greig *et al.*, 1968; Baker, 1973). A facies analysis by the present author (Pauley, 1986, 1990) has resulted in the recognition of four facies associations: turbidite, subaqueous delta, alluvial floodplain, and braided alluvial. These are organised into a broadly upward-coarsening sequence (Figure 2.3). Pyroclastic deposits occur only in the Stretton Group and are not abundant. The main pyroclastic deposits are the Buxton Rock Member of the Burway Formation which is a dust tuff, and the

'Batch Volcanics' of the Synalds Formation which are lapilli and ash tuffs. These latter occur as discrete thick beds within alluvial floodplain facies.

The turbidite facies of the Burway Formation overlie basin-plain mud-

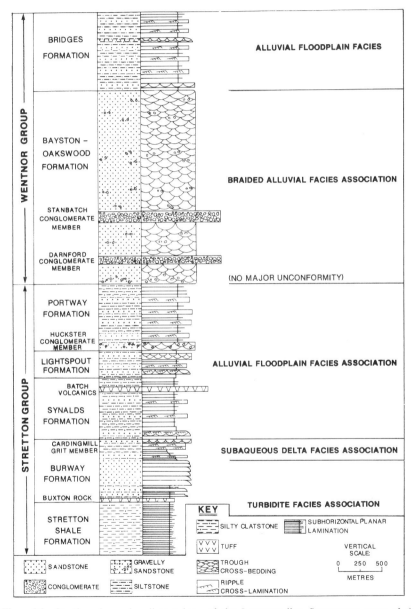

Figure 2.3 Stratigraphy and sedimentology of the Longmyndian Supergroup west of the Church Stretton fault system.

stones of the Stretton Shale Formation. The minimum depth of water in which the latter was deposited is estimated to have been of the order of 700 m. The turbidites were sourced by a fluvially dominated delta which deposited the overlying subaqueous delta facies of the upper part of the Burway Formation. Palaeocurrents within the distributary channel-fills of the delta were towards the northeast to east-northeast. The overlying alluvial flood-plain facies of the Synalds, Lightspout and Portway Formations are inter-preted to have been deposited mainly by sheetfloods and sheetflows which flowed in a westerly to west-northwesterly direction. These facies are interpreted as mainly the distal equivalents of a higher energy braided alluvial system which deposited the overlying sandy and conglomeratic facies of the Bayston-Oakswood Formation. The palaeocurrents which deposited these latter facies were predominantly towards the north-northwest. The facies and palaeocurrent data suggest that a north-northwest-flowing braided fluvial system sourced an alluvial floodplain dominated by west- to west-northwest-flowing sheetfloods and sheetflows. This was crossed by northeast- to east-northeast-flowing fluvial channels which sourced a fluvially dominated delta and a turbidite fan. The west to north-northwest alluvial system may be interpreted as a transverse depositional system sourced from the southeast, whereas the northeast to east-northeast fluvial system may be interpreted as a longitudinal depositional system. These directional data indicate that the basin was elongated in an approximately NE–SW direction. The significance of this is discussed in a later section.

The braided alluvial facies of the Bayston-Oakswood Formation represents the culmination of a progradation of facies in the Stretton Group. A lack of transitional facies beneath the Bayston-Oakswood Formation indicates a rapid influx of coarse detritus that might have been due to a fault-controlled uplift of the source. This is discussed in a later section. The overlying Bridges Formation is composed of alluvial floodplain sediments which are similar to those in the underlying Stretton Group. A return to lower energy conditions might have been due to the denudation of the source, as a result of which fluvial sediments might have onlapped onto the magmatic arc source. Such sediments may be represented by the Willstone Hill Conglomerate which is interpreted to overlie the Uriconian Volcanic Complex unconformably.

2.3   Old Radnor Inlier, Welsh Borderland

The Old Radnor Inlier crops out within the Church Stretton fault system, approximately 40 km to the southwest of the Long Mynd Inlier. It is about 3 km by 1 km in extent and is elongated in a NE–SW direction (Figure 2.4). Precambrian igneous rocks, known as the Stanner-Hanter Complex, crop out to the southeast within, or in proximity to, the Church Stretton fault system

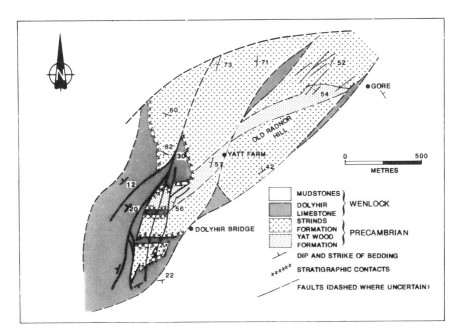

Figure 2.4   Geological map of the Old Radnor Inlier (redrawn after Woodcock, 1988).

(Figure 2.1). Dunning (1975) recognised two fault-bounded lenses of igneous rocks, each about 2 km in length. This type of occurrence is similar to that of the Uriconian Volcanic Complex. Thorpe (1982) thought that the Stanner-Hanter Complex might represent part of the Uriconian volcanic province. The structure of the Old Radnor Inlier has been described by Woodcock (1988), and its stratigraphy and sedimentology by Woodcock and Pauley (1989). The following account is based on these authors.

The Precambrian of the Old Radnor Inlier comprises the Yat Wood Formation and the Strinds Formation which are only seen in fault contact. The Yat Wood Formation, which is at least 90 m thick, comprises pale green siltstones, thinly laminated grey mudstones and interbedded fine grained sandstones. The thinly laminated mudstones contain filamentous and sphaeromorph microfossils (C. Peat, pers. commun., 1987) which are classed as cryptarchs (Peat, 1984a, b). Similar cryptarchs occur within a similar facies in the Lightspout Formation of the Longmyndian Supergroup (Peat, 1984a) and in the Synalds and Portway Formations (Pauley, 1986). In the Longmyndian Supergroup these particular cryptarchs have been found only in the alluvial floodplain facies. On this basis, and on the basis of lithology, the Yat Wood Formation probably correlates with the alluvial floodplain facies of the Longmyndian Supergroup, in particular with the Synalds or Lightspout Formations. An age equivalence cannot be proved by the microfossil evidence although it is probable. The Strinds Formation is at least 600 m thick and comprises fine- to medium-grained sandstone with occasional beds of conglomerate or pebbly sandstone. It is very similar to the Bayston-Oakswood Formation of the Longmyndian Supergroup and is similarly interpreted to have been deposited by braided rivers on an alluvial braidplain.

The sandstones of the Yat Wood and Strinds Formation are petrographically and generally similar to those of the Longmyndian Supergroup. They are predominantly volcaniclastic and, in common with the type Longmyndian, the sediments are interpreted to have been derived directly or indirectly from a magmatic arc. Minor quantities of schist and garnet occur in the Old Radnor sediments, as in the Longmyndian Supergroup, which indicates that metamorphic rocks were present in the source area. The Rushton Schist and Primrose Hill metamorphic rocks, which crop out near Shrewsbury (Figure 2.1), may be fragments of this metamorphic source.

The Precambrian sediments dip moderately to steeply, predominantly towards the west-northwest to northwest. They are unconformably overlain by the Dolyhir Limestone of Wenlock age and are faulted against, or in places overlain by, Wenlock mudstones. The Precambrian and Wenlock sediments are cut by faults which have a predominant strike-slip component and which form a strike-slip duplex. The faults preferentially strike north or north-northeast, and northwest or north-northwest. This fault pattern is interpreted as a Riedel shear response to sinistral strike-slip parallel to the NE–SW trend of the inlier. A pre-Wenlock deformation event is represented by tilting of the Precambrian prior to deposition of the Dolyhir Limestone and by some faults which do not cut the unconformity at the base of the Dolyhir Limestone. This deformation is possibly of Ashgill or late Precambrian–early Cambrian age. Post-Wenlock, probably early Devonian (Acadian), strike-slip faulting appears to be predominant. Later Variscan and Mesozoic fault movements cannot be ruled out.

The Precambrian sediments of the Old Radnor Inlier are closely comparable to those of the Longmyndian Supergroup and are considered to be representatives of it. In both the Long Mynd and Old Radnor Inliers there are fault-bounded associations of Bayston-Oakswood Formation, or its equivalent, with lower stratigraphic horizons: the Stretton Shale Formation within the Church Stretton fault system near Horderley in the Long Mynd Inlier, and the probable equivalent of the Synalds or Lightspout Formation in the Old Radnor Inlier. Both areas also have fault-bounded slivers of Precambrian igneous rocks: the Uriconian Volcanic Complex and its assumed equivalent the Stanner-Hanter Complex. Both areas have been affected by strike-slip faulting along the Church Stretton fault system. These fault-bounded associations are considered to be the result of dissection of the Longmyndian basin and its igneous source during pre-Wenlock, probably late Precambrian, strike-slip faulting.

2.4 Charnian Supergroup, Leicestershire

The Charnian Supergroup crops out in Charnwood Forest, Leicestershire, the inlier being about 10 km by 11 km in extent (Figure 2.5). It consists of approximately 3500 m of tuffs, pelites and greywackes, with subordinate slump breccias, volcanic breccias, conglomerates and quartz arenites. Volcanic rocks of the Whitwick and Bardon Hill Complexes are associated and the Charnian is intruded by, or faulted against, diorites (the Northern and Southern Diorites). The stratigraphy of the Charnian Supergroup has

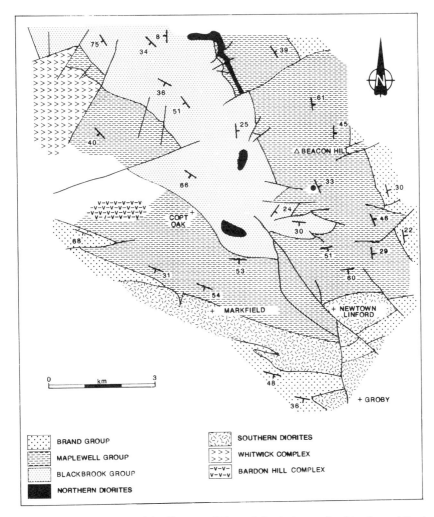

Figure 2.5 Geological map of the Charnwood Forest Inlier (redrawn after Moseley and Ford, 1985).

been revised by Moseley and Ford (1985) and its sedimentology by Moseley and Ford (1989). Much of the following is based on these authors and on Moseley (1979).

The Charnian Supergroup differs from the Longmyndian Supergroup in that there are significant quantities of primary pyroclastics within the Blackbrook and Maplewell Groups. Some of the pyroclastic deposits are of restricted extent and were possibly developed in proximity to con-temporaneous volcanic centres. Remnants of such volcanic centres, the Whitwick and Bardon Hill Complexes, may be contemporaneous with part of the clastic Maplewell Group. The complexes are composed of volcanic breccias, tuffs, and acid to intermediate quartz-feldspar porphyries, which are extrusive in the Whitwick Complex and partly intrusive in the Bardon Hill Complex. Contemporaneous volcanicity in the Longmyndian Supergroup is represented by only occasional pyroclastic deposits, mainly of lapilli or ash tuffs, particularly the 'Batch Volcanics' of the Synalds Formation (Greig *et al.*, 1968; Pauley, 1986). The Longmyndian Supergroup lacks volcanic breccias and agglomerates, debris flows, pull-apart breccias, slumps, and acid to intermediate rocks, all of which are evident in the Charnian Supergroup. Therefore it is concluded that the Charnian Supergroup was deposited closer to, and was more influenced by, contemporaneous volcanic activity than the Longmyndian Supergroup.

The Blackbrook and Maplewell Groups were deposited by a variety of processes which include submarine pyroclastic flows, turbidity currents, and settling from suspensions resulting from pyroclastic and gravity flow mechanisms. These groups are broadly similar to the lower part of the Stretton Group of the Longmyndian Supergroup in that the Stretton Shale Formation comprises basinal shales and the lower and middle parts of the Burway Formation comprise turbidite facies (Pauley, 1990). However, this is an uncertain basis for a stratigraphic correlation since the Longmyndian and Charnian turbidite basins could have developed and been infilled at different times. No unique stratigraphic horizon has been found which is common to both areas and which could be used for correlation purposes. The lower part of the Brand Group has been assigned to a possible littoral/inner neritic environment. If this is correct then an equivalence of facies can be made with the upper part of the Burway Formation of the Longmyndian Supergroup.

An Ediacaran fauna occurs within the Hallgate Member of the Bradgate Formation of the Charnian Supergroup (Ford, 1958, 1963, 1968; Boynton, 1978; Boynton and Ford, 1979). This has not been found in the Longmyndian Supergroup despite the presence of a variety of marine environments. However, possible biogenic structures are present in the shallow marine, low energy sediments at the top of the Burway Formation (Bland, 1984; Pauley, 1986). Some of these were considered to be inorganic by Greig *et al.* (1968). If these structures are biogenic then they are smaller and less complex than the Ediacaran fauna in the Charnian Supergroup or the Llangynog Inlier. The latter is described by Cope (1977, 1979, 1982a). Bland (1984) referred some markings from the Longmyndian to the genus *Arumberia* of Glaessner and Walter (1975), although Pauley (1986) disputed this interpretation. It is possible that the Longmyndian dubiofossils are representative of an earlier, more primitive fauna. This hypothesis is one basis for a proposed correlation between the Longmyndian and Charnian (Figure 2.7). Alternatively they could be the result of different environmental conditions operating at the same time at which the Ediacaran fauna flourished in the Charnian and Llangynog Inliers.

The Longmyndian and Charnian sediments show some common characteristics which are another basis for a proposed correlation (Figure 2.7). The Brand Group of the Charnian Supergroup is more coarse grained than the underlying Hallgate Member of the Bradgate Formation and contains quartz arenites. At its base tuffaceous conglomerates and conglomeratic greywackes have erosive and partly channelised contacts with the Hallgate Member. In the Brand Group occasional pyroclastic deposits are restricted to the basal Hanging Rocks Conglomerate Member. In contrast pyroclastic deposits are relatively abundant in the Blackbrook and Maplewell Groups. It appears that a decline and eventual cessation of volcanicity at the base of the Brand Group was accompanied by a rapid increase in coarse detritus and an increase in the proportion of quartzose clasts. In the Longmyndian Supergroup volcanicity appears to have ceased prior to deposition of the Wentnor Group which also represents a rapid influx of coarse detritus. A petrographic analysis of the Longmyndian sediments (Pauley, 1986) revealed that the Wentnor Group is more variable in composition with respect to the Stretton Group and several samples contained a distinctly higher proportion of quartz and feldspar. Consequently there are some similarities between the Charnian Supergroup and the Longmyndian Supergroup in these respects, which suggest that the Brand Group may be equivalent in time to the Wentnor Group and the Stretton Group may be equivalent in time to the Blackbrook and Maplewell Groups. This statement is based on a proposed regional tectonosedimentary mechanism which is discussed in a following section.

The Charnian and Longmyndian sediments are similar in that they are predominantly volcaniclastic and were sourced by a magmatic arc or arcs. Moseley and Ford (1989) proposed that the Charnian accumulated in a trough with an approximate NE–SW trend. This trend is similar to that proposed for the Longmyndian basin. One difference between the Charnian and Longmyndian basins is that in the Charnian there is evidence for a palaeoslope towards the southeast whereas in the Longmyndian the palaeocurrent data indicate derivation from the southeast, in addition to northeast to east-northeast longitudinal flow. The significance of the palaeocurrent data and the palaeoslope indicators is discussed in a following section.

The Northern and Southern Diorites intrude, or are faulted against, the

Charnian Supergroup. These have yielded radiometric age-dates of 304 ± 90 Ma, and 540 ± 57 Ma, respectively (Pankhurst, 1982, after Cribb, 1975). The diorites were emplaced prior to the development of cleavage (Boulter and Yates, 1987). Pharaoh *et al.* (1987a) argued on the basis of mica crystallinity data that the Charnian cleavage is of late Precambrian age. These data indicate that the Charnian metasediments are of a higher metamorphic grade (metapelite stage III) than the overlying Cambrian sediments (metapelite stage I). The 540 ± 57 Ma age-date may represent the time at which the Charnian Supergroup was cleaved and deformed. The younger Variscan date may be the result of a much later phase of deformation. Felsic volcanics of the Bardon Hill and Whitwick Complexes have yielded a Rb-Sr age-date of 512 ± 34 Ma (Pharaoh *et al.*, 1987a). This age-date and the 540 ± 57 Ma age-date are similar to a number of radiometric age-dates of between *c.* 520 Ma and *c.* 535 Ma obtained from the Precambrian of the Long Mynd area which are thought to be the result of a tectonic event which probably involved sinistral transpression (Pauley, 1990). This suggests that the Charnian Supergroup was deformed at about the same time as the Longmyndian Supergroup. No firm evidence for strike-slip deformation of the Charnian Supergroup has been documented. However Pharaoh *et al.* (1987a) suggested that the anticlockwise transecting cleavage noted by Evans (1979) is compatible with dextral transpression. Since the cleavage is thought to have been imposed in late Precambrian time then the Charnian might have been involved in transpression during the late Precambrian.

2.5 Llangynog Inlier, Dyfed

Small outcrops of Precambrian strata occur within an inlier of mainly Cambrian to Arenig rocks in the vicinity of Llangynog, near Carmarthen, South Wales (Figure 2.6). Cope (1982b) referred to this inlier as the Llangynog Inlier. The Precambrian rocks and the Ediacaran fauna which they contain have been described by Cope (1977, 1979, 1982b). His accounts form the basis of much of the following discussion.

Two rock groups are recognised: rhyolitic and andesitic lavas and tuffs with intrusions of dolerite; and interlaminated siltstones and fine grained sandstones which appear to be interbedded with the igneous rocks. The sediments have yielded an Ediacaran fauna which includes *Cyclomedusa* sp., *Medusinites* sp., shallow branching burrow systems, and feeding trails. Based on the presence of this fauna Cope (1982b) suggested an age for the sediments in the range of *c.* 680–580 Ma. The sediments may be correlated broadly with the Charnian Supergroup based on the presence of an Ediacaran fauna in both. A shallow marine origin for the sediments has been proposed and the sediments are comparable, though not necessarily equivalent to, similar shallow marine sediments at the top of the Burway Formation of the

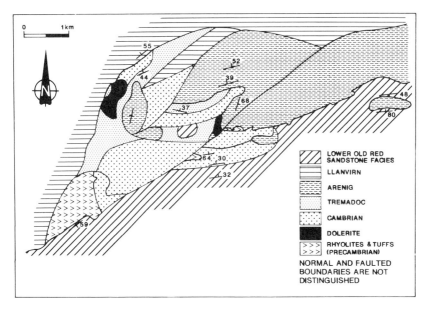

Figure 2.6   Geological map of the Llangynog Inlier (redrawn and adapted from Cope, 1982, with permission of the National Museum of Wales).

Longmyndian Supergroup. In common with the Charnian and Longmyndian, the sediments of the Llangynog Inlier are composed mainly of volcanic detritus. A similar biogenic structure to the possibly biogenic 'small ring-like feature' of Pauley (1986), which occurs within the upper part of the Burway Formation, has been recognised in the Llangynog sediments by the present author. This suggests the possibility of a broad correlation. However an Ediacaran fauna of the type seen in the Llangynog Inlier has not been found in the Longmyndian. It is possible that the dubiofossils at the top of the Burway Formation of the Longmyndian Supergroup are representative of an earlier, more primitive fauna than that seen in the Llangynog Inlier. This is one basis for a proposed correlation (Figure 2.7).

Cope (1979) favoured a thrust fault origin for the structure of the inlier. The inlier lies to the north of the Llandyfaelog disturbance which may be a continuation of the Church Stretton fault system.

2.6   Other outcrops of Precambrian sediments in England and the Welsh Borderland

A small outcrop of Precambrian sediments is reported to occur at Pedwardine, near Brampton Bryan in the vicinity of the Church Stretton fault

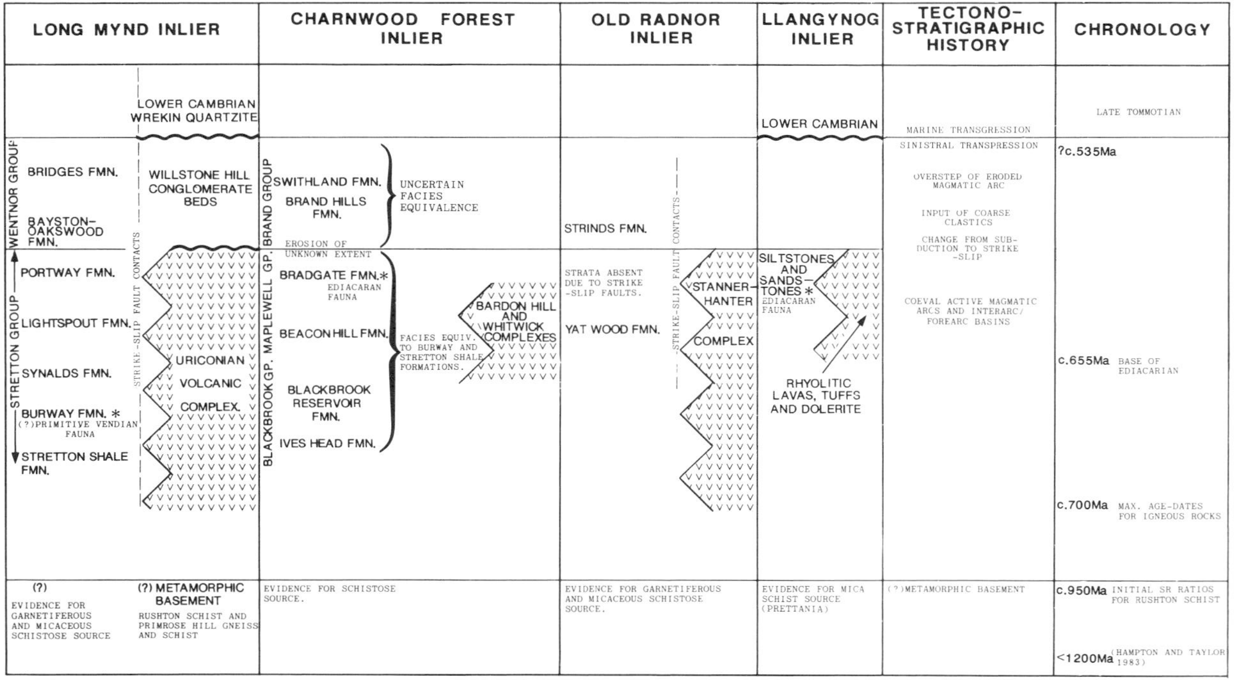

Figure 2.7   A suggested correlation of the Longmyndian Supergroup and related Precambrian strata of England and Wales.

system, south-southwest of the Long Mynd (Dunning, 1975). These sediments have been correlated with the Bayston-Oakswood Formation (Cox, 1912). An inlier of Precambrian rocks also occurs to the south of the Malvern Hills near Huntley, Gloucestershire. These were correlated with the Longmyndian Supergroup by Callaway (1900). However Ziegler et al. (1968) recognised volcanic flows with interbedded red beds which they correlated with the Malvernian.

2.7   Depositional setting of the Precambrian sediments

The sedimentary sequences reviewed in the previous account are all associated with Precambrian igneous rocks. Thorpe et al. (1984) considered that these are representative of volcanic arcs which were the result of south-easterly subduction beneath southern Britain. Pharaoh et al. (1987b) argued, from geochemical evidence, that the Charnian volcanic suite was produced in a calc-alkaline volcanic arc located on either oceanic or on juvenile continental crust. The Uriconian lavas have a component of arc magmatism but have characteristics transitional between within-plate and volcanic arc lavas (Thorpe et al., 1984; Pharaoh et al., 1987b). Pharaoh et al. (1987b) considered that they could have been erupted in a fault-controlled ensialic marginal basin. Thorpe (1972, 1974, 1979, 1982) argued that the Uriconian Volcanic Complex represents a continental margin volcanic arc. The Warren House group, which is in tectonic contact with the Malvernian plutonic complex, has the geochemical characteristics of marginal basin and island arc tholeiites (Thorpe et al., 1984; Pharaoh et al., 1987b). The Malvernian plutonic complex is of calc-alkaline character and has characteristics of volcanic arc rocks (Thorpe et al., 1984). Pharaoh et al. (1987b) argued that the geochemical dissimilarity between the Uriconian and Charnian igneous rocks justifies their distinction as terranes, the Malvern Lineament representing the suture between them.

The sediments reviewed in this account have all been derived in the main from volcanic and plutonic rocks of which the aforementioned igneous rocks are considered to be representatives. The lithologies of the clasts within the sediments can be directly matched with some of the Precambrian igneous rocks. There is evidence that sedimentation was, at least partly, coeval with volcanic activity. QFL plots for the sediments of the Longmyndian Supergroup (Pauley, 1986) show that the Longmyndian Supergroup plots close to the field of undissected magmatic arc provenances as defined by Dickinson and Suczek (1979). The majority of the Longmyndian plots close to the field of forearc basin sediments as defined by Valloni and Maynard (1981). In addition, the preponderance of glassy volcanic clasts in the Longmyndian Supergroup is a characteristic of forearc basin sediments (Valloni and Maynard, 1981; Maynard et al., 1982). Southeasterly subduction (Thorpe et *al.*, 1984), a southeasterly magmatic arc source for the Longmyndian, and the evidence of probable subduction in the Mona Complex (Gibbons, 1984; Thorpe *et al.*, 1984) also suggest that the Longmyndian was deposited in a forearc basin. Forearc basin sediments typically show upward-shallowing trends, typically passing upwards from turbidite and associated facies, to deltaic or shelf deposits, and finally to terrigenous deposits (Dickinson and Seely, 1979). The Longmyndian Supergroup is similar in this respect. Within forearc basins longitudinal sediment dispersal can occur as a result of uplift of the subduction complex which restricts the width of the basin. The ENE–NE-flowing distributary channels of the Longmyndian Supergroup are thought to represent such longitudinal sediment dispersal.

The Longmyndian Supergroup is similar to the forearc basin deposits of the Great Valley sequence of California. This sequence rests on oceanic crust and comprises a very thick sequence of turbidite deposits, which pass upwards into deltaic deposits, followed by nonmarine deposits (Dickinson and Seely, 1979). The sequence has been deformed into a large syncline, up to about 15 km deep, which is comparable in magnitude to the Longmyndian syncline. The Great Valley sequence accumulated over a period of time similar to that proposed for deposition of the Longmyndian Supergroup; a maximum of about 165 Ma. This value represents the time between the maximum radiometric age-data for the Precambrian igneous rocks (702 ± 8 Ma for the Stanner-Hanter Complex; Patchett *et al.*, 1980) and the proposed age-date of deformation of the Longmyndian Supergroup (*c.* 535 Ma).

The NE–SW trend of the Charnian basin (Moseley and Ford, 1989) is similar to that proposed for the Longmyndian basin. The trend of these basins is compatible with southeasterly subduction. However Maguire *et al.* (1981) and Le Bas (1982) argued for northeasterly subduction in the English Midlands which is difficult, though not impossible, to reconcile with the palaeocurrent data and configuration of the Charnian basin. The Charnian basin is different from the Longmyndian basin in that volcanic centres are thought to have lain to the northwest (Moseley, 1979; Moseley and Ford, 1989). If southeasterly subduction is accepted then this suggests that the Charnian Supergroup was deposited in a backarc or interarc basin. South-easterly subduction and NE–SW trending basins are compatible in orientation with the NE–SW trending strike-slip faults of the Welsh Borderland Fault System (Woodcock and Gibbons, 1988), some of which cut the Longmyndian Supergroup, and with the NE–SW trending faults, such as the Menai Strait Fault System (Figure 2.1) which cut the Mona Complex (Gibbons, 1987). Pauley (1990) argued for a late Precambrian–early Cambrian sinistral transpressional origin for the Church Stretton and Pontesford-Linley fault systems. Gibbons (1987) and Dallmeyer and Gibbons (1987) argued for a similar origin for the Menai Strait Fault System. Comparable, probably late Precambrian sinistral transcurrent movements are also recognised in the Precambrian Rosslare Complex of southeast Ireland (Murphy, 1990).

Late Precambrian–early Cambrian sinistral transpression is thought to have terminated magmatic arc activity, to have juxtaposed diverse elements of the Precambrian magmatic arc and associated basins across strike-slip faults, and to have deformed the Precambrian prior to erosion and deposition of late Tommotian shallow marine quartzites. Pauley (1990) proposed that radiometric age-dates in the region of c. 535 Ma to c. 520 Ma obtained from the Precambrian of Shropshire are the result of this late Precambrian–early Cambrian deformation. Other radiometric age-dates which are thought to be the result of this deformation are 540 ± 57 Ma for the Southern Diorite (Pankhurst, 1982, after Cribb, 1975), and 512 ± 34 Ma for felsic volcanics of the Bardon Hill and Whitwick Complexes (Pharaoh et al., 1987a). As discussed by Pauley (1990) and Cope and Gibbons (1987), the radiometric age-dates appear to require the base of the Cambrian, in this case, to be younger than is generally accepted, i.e. less than c. 535 Ma. Alternatively the radiometric age-dates may be the result of a tectonic event during the Cambrian. Such tectonic activity is represented, for example, by an unconformity between the Lower and Middle Cambrian of Shropshire (Greig et al., 1968). Since similar radiometric age-dates were obtained by fission-track and Rb-Sr methods, it is unlikely that the age-dates are anomalous.

It is likely that strike-slip faults were in existence during Precambrian subduction considering that non-oblique subduction is a unique case. As discussed previously, the change in Charnian and Longmyndian sedimentation at the base of the Brand Group and Wentnor Group, respectively, may be the result of a change from a subduction-dominated setting to one dominated by strike-slip faulting. Such a change, accompanied by transpression, would have resulted in the recorded cessation of volcanicity, uplift of the magmatic arc, and increase in the supply of coarse detritus with a higher proportion of quartzo-feldspathic plutonic rock fragments. One alternative explanation is that subduction shifted to the northwest, resulting in a cessation of volcanicity in the earlier arc to the southeast. However, this might be expected to have resulted in back-arc extension, there being no evidence for this in the sedimentary record. The initiation of strike-slip faulting towards the end of Charnian and Longmyndian sedimentation is a logical prelude to the proposed late Precambrian–early Cambrian sinistral transpression which terminated late Precambrian magmatic arc activity and associated sedimentation. The mechanism proposed here is one basis for a proposed correlation of the Charnian and Longmyndian sequences (Figure 2.7).

The Charnian, Longmyndian and Old Radnor sediments contain minor quantities of metamorphic rock fragments. The recorded metamorphic detritus includes schistose polycrystalline quartz, quartz-muscovite schist, quartz-muscovite-chlorite schist, and garnet. In some of the schist fragments relict igneous textures can be recognised (Pauley, 1986). Metamorphic rocks which could have supplied such rock fragments are the Rushton Schist and Primrose Hill Gneiss of Shropshire and the metamorphic rocks of the Malvernian plutonic complex. The two former rock types are garnetiferous (Pocock et al., 1938; Pauley, 1986) and some of the Malvernian metamorphic rocks contain garnet (Lambert and Holland, 1971). Initial Sr ratios for the Rushton Schist, which is interpreted as having a sedimentary protolith, suggest a pre-metamorphic age of c. 950 Ma (Thorpe et al., 1984). A Rb-Sr age-date of 667 ± 20 Ma for the same rocks (Thorpe et al., 1984) suggests that metamorphism of the Rushton Schist was coincident with magmatic arc activity. A later age-date of 536 ± 8 Ma (Patchett et al., 1980) is probably due to late Precambrian–early Cambrian deformation. Lambert and Holland (1971) argued that schists in the Malvernian resulted from the shearing and hydrolysis of plutonic igneous rocks, although one garnet-mica schist was suggested to have had a sedimentary protolith. These data suggest that the Precambrian metamorphic rocks which sourced the late Precambrian basins could represent the uplifted metamorphosed roots of the early magmatic arc with its associated sediments. This hypothesis is supported by the maximum model ages for the Rushton Schists of Tchur 1300 Ma, and Tmorb 1600 Ma (Thorpe et al., 1984) and by the Sr and Pb isotopic data of Hampton and Taylor (1983) who argued that the basement in southern Britain is c. 800 Ma and definitely less than 1200 Ma-old. The Rushton Schists and Primrose Hill Gneiss are thought to be slices of a metamorphic source area which has been dissected by the Church Stretton fault system. Cope and Bassett (1987) and Cope (1987) recognised a Precambrian metamorphic landmass ('Pretannia') to the south of South Wales which supplied mica and mica-schist fragments to the Lower Palaeozoic sediments of South Wales. It is possible that Pretannia was not located to the south of South Wales during the late Precambrian considering that it might have been translated during later transcurrent movements. However the magnitude of any displacement of Pretannia with respect to the late Precambrian magmatic arc and its associated basins is conjectural at the present time and Pretannia could have acted as a source for the late Precambrian sediments of England and Wales.

Acknowledgements

Research was carried out during tenure of a Natural Environment Research Council (NERC) grant to the author at Liverpool University under the supervision of Drs P. Brenchley and P. Crimes. Nigel Woodcock, John Cope, Trevor Ford and John Moseley collaborated during writing and provided critiques of the draft paper. Their assistance is gratefully acknowledged. The Geochem Group Limited, Chester, provided technical assistance (in typing and drafting).

References

Baker, J. W. (1973) A marginal Late Proterozoic ocean basin in the Welsh region. Geol. Mag. 110, 447–455.

Bath, A. H. (1974) New isotopic age data on rocks from the Long Mynd, Shropshire. *J. Geol. Soc. London* **130**, 567–574.

Bland, B. H. (1984) *Arumberia* Glaessner and Walter, a review of its potential for correlation in the region of the Precambrian–Cambrian boundary. *Geol. Mag.* **121**, 625–633.

Boulter, C. A. and Yates, M. G. (1987) Confirmation of the pre-cleavage emplacement of both the northern and southern diorites into the Charnian Supergroup. *Mercian Geol.* **10**, 281–286.

Boulton, W. S. (1904) The igneous rocks of Pontesford Hill (Shropshire). *Q. J. Geol. Soc. London* **60**, 450–486.

Boynton, H. E. (1978) Fossils from the Precambrian of Charnwood Forest, Leicestershire. *Mercian Geol.* **6**, 291–296.

Boynton, H. E. and Ford, T. D. (1979) *Pseudovendia charnwoodensis*—a new Precambrian arthropod from Charnwood Forest, Leicestershire. *Mercian Geol.* **7**, 175–177.

Brasier, M. D. (1985) Evolutionary and geological events across the Precambrian–Cambrian boundary. *Geology Today*, Sept.–Oct., 141–146.

Callaway, C. (1891) On the unconformities between the rock systems underlying the Cambrian quartzite in Shropshire. *Q. J. Geol. Soc. London* **47**, 109–125.

Callaway, C. (1900) On the Longmyndian inliers at Old Radnor and Huntley (Gloucestershire). *Q. J. Geol. Soc. London* **83**, 551–573.

Cobbold, E. S. and Whittard, W. F. (1935) The Helmeth Grits of the Caradoc Range, Church Stretton: their bearing on part of the Precambrian succession of Shropshire. *Proc. Geol. Assoc. London* **46**, 348–359.

Cope, J. C. W. (1977) An Ediacara-type fauna from South Wales. *Nature (London)* **268**, 624.

Cope, J. C. W. (1979) Early history of the southern margin of the Twyi anticline in the Carmarthen area, South Wales. In *The Caledonides of the British Isles, Reviewed*, eds. Harris, A. L., Holland, C. H. and Leake, B. E., *Spec. Publ. Geol. Soc. London* **8**, 527–532.

Cope, J. C. W. (1982a) Precambrian fossils of the Carmarthen area, Dyfed. *Nature in Wales*, New Ser. **1**, 11–16.

Cope, J. C. W. (1982b) The geology of the Llanstephan Peninsula. In *Geological Excursions in Dyfed, Southwest Wales*, ed. Bassett, M. G., 259–269.

Cope, J. C. W. (1987) The Pre-Devonian geology of southwest England. *Proc. Ussher Soc.* **7**, 468–473.

Cope, J. C. W. and Bassett, M. G. (1987) Sediment sources and Palaeozoic history of the Bristol Channel area. *Proc. Geol. Assoc. London* **98**, 315–330.

Cope, J. C. W. and Gibbons, W. (1987) New evidence for the relative age of the Ercall Granophyre and its bearing on the Precambrian–Cambrian boundary in southern Britain. *Geol. J.* **22**, 53–60.

Cox, A. H. (1912) On an inlier of Longmyndian and Cambrian rocks at Pedwardine (Herefordshire). *Q. J. Geol. Soc. London* **68**, 364–373.

Cribb, S. J. (1975) Rubidium–strontium ages and strontium isotope ages from the igneous rocks of Leicestershire. *J. Geol. Soc. London* **131**, 203–212.

Dallmeyer, R. D. and Gibbons, W. (1987) The age of blueschist metamorphism in Anglesey, North Wales: evidence from $^{40}Ar/^{39}Ar$ mineral dates of the Penmynydd schists. *J. Geol. Soc. London* **144**, 843–852.

Dean, W. T. (1964) The geology of the Ordovician and adjacent strata in the southern Caradoc district of Shropshire. *Bull. Br. Mus. (Nat. Hist.), Geol.* **9**, 259–296.

Dickinson, W. R. and Seely, D. R. (1979) Structure and stratigraphy of forearc regions. *Bull. Am. Assoc. Pet. Geol.* **63**, 2–31.

Dickinson, W. R. and Suczek, C. A. (1979) Plate tectonics and sandstone compositions. *Bull. Am. Assoc. Pet. Geol.* **63**, 2164–2182.

Dunning, F. W. (1975) Precambrian craton of central England and the Welsh Borders. In *A Correlation of Precambrian Rocks in the British Isles*, eds. Harris, A. L., Shackleton, R. M., Watson, J., Downie, C., Harland, W. B. and Moorbath, S., *Spec. Rep. Geol. Soc. London* **6**, 83–95.

Evans, A. M. (1979) The East Midlands aulacogen of Caledonian age. *Mercian Geol.* **7**, 31–42.

Ford, T. D. (1958) Precambrian fossils from Charnwood Forest. *Proc. Yorks. Geol. Soc.* **31**, 211–217.

Ford, T. D. (1963) The Precambrian fossils of Charnwood Forest. *Trans. Leics. Lit. Phil. Soc.* **57**, 57–62.

Ford, T. D. (1968) Pre-Cambrian palaeontology of Charnwood Forest. In *Geology of the East Midlands*, eds. Sylvester-Bradley, P. C. and Ford, T. D., Univ. Leicester Press, 400 pp.

Gibbons, W. (1984) The Precambrian basement of England and Wales. *Proc. Geol. Assoc. London* **95**, 387–389.

Gibbons, W. (1987) Menai Strait fault system: an early Caledonian terrane boundary in North Wales. *Geology* **15**, 744–747.

Glaessner, M. F. and Walter, M. R. (1975) New Precambrian fossils from the Arumbera Sandstone, Northern Territory, Australia. *Alcheringa* **1**, 59–69.

Greig, D. C., Wright, J. E., Hains, B. A. and Mitchell, H. G. (1968) *Geology of the Country Around Church Stretton, Craven Arms, Wenlock Edge and Brown Clee*. Mem. Geol. Surv. G.B., HMSO London, 379 pp.

Hampton, C. M. and Taylor, P. N. (1983) The age and nature of the basement of southern Britain: evidence from Sr and Pb isotopes in granites. *J. Geol. Soc. London* **140**, 499–509.

James, J. H. (1952a) Notes on the relationship of the Uriconian and Longmyndian rocks, near Linley, Shropshire. *Proc. Geol. Assoc. London* **63**, 198–200.

James, J. H. (1952b) The structure and stratigraphy of the Precambrian of the Longmynd Area, Shropshire. Unpublished Ph.D. thesis, Bristol University, 95 pp.

James, J. H. (1956) The structure and stratigraphy of part of the Precambrian outcrop between Church Stretton and Linley, Shropshire. *Q. J. Geol. Soc. London* **112**, 315–337.

Lambert, R. St. J. and Holland, J. G. (1971) The petrography and chemistry of the igneous complex of the Malvern Hills, England. *Proc. Geol. Assoc. London* **82**, 323–352.

Lapworth, C. and Watts, W. W. (1910) Geology in the field: Shropshire. *Proc. Geol. Assoc. London*, Jubilee Vol., 747–749.

Le Bas, M. J. (1982) Geological evidence from Leicestershire on the crust of southern Britain. *Trans. Leics. Lit. Phil. Soc.* **76**, 54–67.

Lynas, B. D. T. (1988) Evidence for strike-slip faulting in the Shelve Ordovician inlier, Welsh Borderland: implications for the south British Caledonides. *Geol. J.* **23**, 39–57.

Maguire, P. K. H., Whitcombe, D. N. and Francis, D. J. (1981) Seismic studies in the Central Midlands of England 1975–1980. *Trans. Leics. Lit. Phil. Soc.* **75**, 58–66.

Maynard, J. B., Valloni, R. and Ho-Shing, Yu. (1982) Composition of modern deep-sea sands from arc-related basins. In *Trench-Forearc Geology: Sedimentation and Tectonics on Modern and Ancient Active Plate Margins*, ed. Leggett, J. K. *Spec. Publ. Geol. Soc. London* **10**, 551–561.

Moseley, J. (1979) The geology of the Late Precambrian rocks of Charnwood Forest, Leicestershire. Unpublished Ph.D. thesis, Leicester University.

Moseley, J. and Ford, T. D. (1985) A stratigraphic revision of the late Precambrian rocks of Charnwood Forest, Leicestershire. *Mercian Geol.* **10**, 1–18.

Moseley, J. and Ford, T. D. (1989) The sedimentology of the Charnian Supergroup. *Mercian Geol.* **11**, 251–274.

Murphy, F. C. (1990) Basement-cover relationships of a reactivated Cadomian mylonite zone: Rosslare Complex, S.E. Ireland. In *The Cadomian Orogeny*, eds. D'Lemos, R. S., Strachan, R. A. and Topley, C. G., *Spec. Publ. Geol. Soc. London* **51**, 329–339.

Naeser, C. W., Toghill, P. and Ross, R. J. (1982) Fission-track ages from the Precambrian of Shropshire. *Geol. Mag.* **119**, 213–214.

Pankhurst, R. J. (1982) Geochronilogical tables for British igneous rocks. Appendix C. In *Igneous Rocks of the British Isles*, ed. Sutherland, D. S., Wiley, Chichester, 575–582.

Patchett, P. J., Gale, N. H., Goodwin, R. and Humm, M. J. (1980) Rb–Sr whole-rock isochron ages of late Precambrian to Cambrian igneous rocks from southern Britain. *J. Geol. Soc. London* **137**, 649–656.

Pauley, J. C. (1986) The Longmyndian Supergroup: facies, stratigraphy and structure. Unpublished Ph.D. thesis, University of Liverpool, 360 pp.

Pauley, J. C. (1990) Sedimentology, structural evolution and tectonic setting of the late Precambrian Longmyndian Supergroup of the Welsh Borderland, UK. In *The Cadomian Orogeny*, eds. D'Lemos, R. S., Strachan, R. A. and Topley, C. G., *Spec. Publ. Geol. Soc. London* **51**,

Peat, C. J. (1984a) Precambrian microfossils from the Longmyndian of Shropshire. *Proc. Geol. Assoc. London* **95**, 17–22.

Peat, C. J. (1984b) Comments on some of Britain's oldest microfossils. *J. Micropalaeontol.* **3**, 65–71.

Pharaoh, T. C., Merriman, R. J., Webb, P. C. and Beckinsale, R. D. (1987a) The concealed Caledonides of eastern England: preliminary results of a multidisciplinary study. *Proc. Yorks. Geol. Soc.* **46**, 355–369.

Pharaoh, T. C., Webb, P. C., Thorpe, R. S. and Beckinsale, R. D. (1987b) Geochemical evidence for the tectonic setting of late Proterozoic volcanic suites in central England. In *Geochemistry and Mineralization of Proterozoic Volcanic Suites*, eds. Pharaoh, T. C., Beckinsale, R. D. and Rickard, D., *Spec. Publ. Geol. Soc. London* **33**, 541–552.

Pocock, R. W., Whitehead, T. H., Wedd, C. B. and Robertson, T. (1938) *Geology of the Shrewsbury District*. Mem. Geol. Surv. G.B., HMSO London, 297 pp.

Taylor, J. H. (1958) Precambrian sedimentation in England and Wales. *Eclog. Geol. Helv.* **51**, 1078–1092.

Thorpe, R. S. (1972) The geochemistry and correlation of the Warren House, the Uriconian and the Charnian volcanic rocks from the English Precambrian. *Proc. Geol. Assoc. London* **83**, 269–285.

Thorpe, R. S. (1974) Aspects of magmatism and plate tectonics in the Precambrian of England and Wales. *Geol. J.* **9**, 115–136.

Thorpe, R. S. (1979) Late Precambrian igneous activity in S. Britain. In *The Caledonides of the British Isles: Reviewed*, eds. Harris, A. L., Holland, C. H. and Leake, B. E., *Spec. Publ. Geol. Soc. London* **8**, 579–584.

Thorpe, R. S. (1982) Precambrian igneous rocks of England, Wales and southeastern Ireland. In *Igneous Rocks of the British Isles*, ed. Sutherland, D. S., Wiley, 19–35.

Thorpe, R. S., Beckinsale, R. D., Patchett, P. J., Piper, J. D. A., Davies, G. R. and Evans, J. A. (1984) Crustal growth and late Precambrian–early Palaeozoic plate tectonic evolution of England and Wales. *J. Geol. Soc. London* **141**, 521–536.

Valloni, R. and Maynard, J. B. (1981) Detrital modes of recent deep-sea sands and their relation to tectonic setting: a first approximation. *Sedimentology* **28**, 75–83.

Woodcock, N. H. (1984a) Early Palaeozoic sedimentation and tectonics in Wales. *Proc. Geol. Assoc. London* **95**, 323–335.

Woodcock, N. H. (1984b) The Pontesford Lineament, Welsh Borderland. *J. Geol. Soc. London* **141**, 1001–1014.

Woodcock, N. H. (1988) Strike-slip faulting along the Church Stretton Lineament, Old Radnor Inlier, Wales. *J. Geol. Soc. London* **145**, 925–933.

Woodcock, N. H. and Gibbons, W. (1988) Is the Welsh Borderland fault system a terrane boundary? *J. Geol. Soc. London* **145**, 915–923.

Woodcock, N. H. and Pauley, J. C. (1989) The Longmyndian rocks of the Old Radnor Inlier, Welsh Borderland. *Geol. J.* **24**, 113–120.

Woodcock, N. H., Awan, M. A., Johnson, T. E., Mackie, A. H. and Smith, R. D. A. (1988) Acadian tectonics of Wales during Avalonia/Laurentia convergence. *Tectonics* **7**, 483–495.

Ziegler, A. M., Cocks, L. R. M. and McKerrow, W. S. (1968) The Llandovery transgression of the Welsh Borderland. *Palaeontology* **11**, 736–782.

3 Pre-Arenig terranes of northwest Wales

WES GIBBONS

3.1 Introduction and historical perspective

'Anglesea is almost as distinct in structure from Snowdonia, as if they had been separated by the Atlantic sea rather than the Straits of Menai', Adam Sedgwick (in a letter to Murchison, dated 13 September 1831).

The geology of northwest Wales is notable for the exposure of a wide range of igneous and metamorphic rocks beneath a transgressive Arenig overstep sequence. This sub-Arenig basement is widely exposed northwest of the Menai Strait across the island of Anglesey (Ynys Môn) and on the mainland along the northwest side of the Llŷn peninsula (Figure 3.1). Blake (1888) referred to these rocks as belonging to the *Monian System*, a term superceded by Greenly's (1919) *Mona Complex*. More recently, it has been recognised that several disparate, and possibly unrelated geological units occur within the area. On the Llŷn peninsula, for example, Monian melange is separated by a steep shear zone from a series of mostly granitic dioritic and gneissic rocks now referred to as the *Sarn Complex* (Gibbons, 1983a). Similarly, in central Anglesey, a strip of mostly gneisses and granites lies in tectonic contact with adjacent rocks, and has been defined as belonging to a *Coedana Complex* (after the Coedana Granite) (Bassett *et al.*, 1986). Thus Greenly's Mona Complex may be better viewed as including more than one fault-isolated igneous and metamorphic complex so that *Monian Supercomplex* becomes a more appropriate collective term for the entire outcrop of these rocks.

The history of research on the Monian Supercomplex has been dealt with in Gibbons (1983b) and will not be repeated in detail here. In Anglesey a major phase of 19th century work by several authors, notably Blake and Matley, culminated in the early 20th century with a detailed survey by Greenly (1919). At about the same time Matley (1913, 1928) mapped the Monian rocks on Llŷn in similar detail, although unfortunately he published relatively little, with the result that this area subsequently remained poorly known in comparison with Anglesey. The first important revision of Greenly's work was effected by Shackleton (1954) who replaced Greenly as the major figure in Anglesey geology (Shackleton, 1969, 1975). With the advent of plate tectonic theory, the potential significance of Monian blue-schists (first described by Blake, 1888), melanges, serpentinites and other

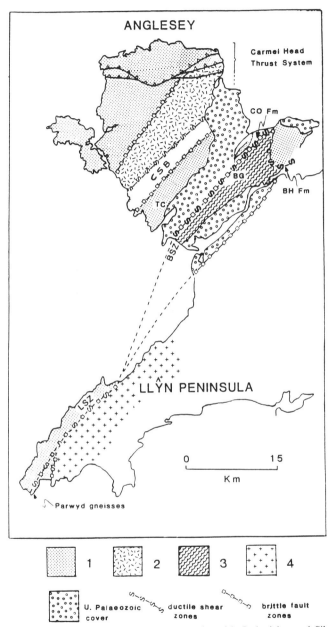

Figure 3.1   The sub-Arenig geology of northwest Wales with Ordovician and Silurian cover removed. 1 Monian Supergroup; 2 Coedana Complex; 3 Blueschist belt (Eastern Schist Belt of Figure 3.2); 4 Sarn Complex. CSB = Central schist belt involving both Coedana Complex and Monian Supergroup, BSZ = Berw shear zone, LSZ = Llyn shear zone, CO Fm = Careg Onen Formation, BH Fm = Baron Hill Formation, BG = Bwlch Gwyn Tuff, TC = Trefdraeth Conglomerate. The Carmel Head Thrust System is an Acadian structure, latest movements of which affect pre-Carboniferous red beds interpreted as Devonian in age.

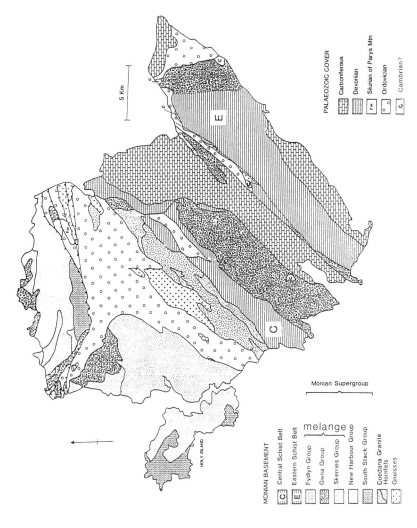

Figure 3.2   Geological map of Anglesey (based on Greenly, 1919 and Shackleton, 1975).

lithologies was recognised by several authors, notably Dewey (1969) and Wood (1974). The Mona Complex came to be viewed as recording the presence of a late Precambrian subduction system, and was incorporated into various models that involved the consumption of oceanic crust, usually beneath southern Britain to the southeast (see references in Gibbons, 1983b).

Shackleton interpreted the typically narrow transitions between rocks of differing metamorphic grade at various localities on Llŷn and Anglesey as due to rapid prograde metamorphism. A major revision of this interpretation was instigated by Baker (1969) who recognised that some of the key sections on which the Shackleton model of 'prograde metamorphic transition' were based were actually mylonitic shear zones juxtaposing high and low grade rocks. This alternative interpretation led to a polarisation of views and some confusion was generated within subsequent literature. Those subscribing to the prograde transition theory viewed the gneisses and granitic rocks (such as those in the Sarn Complex and in central Anglesey) as derived from lower grade Monian metasediments, whereas Baker's shear zone interpretation removed the need for a known relative age relationship. The rocks in question belong to what Greenly (1919) had referred to as the *Penmynydd Zone of Metamorphism* and form three NE–SW striking metamorphic belts in central and southeast Anglesey (Figure 3.2) and on the Llŷn peninsula. Following the work of Barber and Max (1979), Gibbons (1981, 1987a) and Mann (1986), many of Greenly's 'Penmynydd' lithologies have been recognised to be ductile fault rocks, although detailed work on these rocks has only yet been published for the Llŷn exposures (Gibbons, 1983a).

A controversial publication by Barber and Max in 1979 challenged previously held views and provoked considerable argument (see discussion in Barber and Max, 1979). As well as favouring Baker's shear zone interpretation, these authors went much further by suggesting radical reinterpretations of Shackleton's Monian stratigraphy (revised after Greenly). Furthermore, these authors argued for a Cambrian, rather than Precambrian age for part of the supercomplex. These subjects are discussed in more detail later in this chapter.

The increasing recognition of ductile fault rocks on Anglesey and Llŷn, often occurring in subvertical shear zones that separate radically different rock units, led directly to the application of the suspect terrane concept for this area (Gibbons, 1983b, 1987a, 1989). The model has emphasised the presence of sinistral, transcurrent kinematic indicators in one of the most prominent of these shear zone boundaries (the Berw shear zone in southeast Anglesey; Figure 3.1). According to this model, the whole outcrop is viewed as comprising at least four suspect terranes, the boundaries to all of which are tectonic (Figure 3.3). These four terranes are the Sarn Complex, the blueschists in southeast Anglesey, the Coedana Complex, and a thick sequence of low grade metasediments and melange called the Monian Supergroup (after Shackleton, 1975; previously the 'Bedded Succession' of

Figure 3.3   Cartoon depicting the application of the suspect terrane concept to northwest Wales. 1 Monian Supergroup; 2 Coedana Complex; 3 Blueschist belt; 4 Sarn Complex.

Greenly, 1919). The Sarn Complex is viewed as the likely northwestern edge of the 'Avalonian' basement seen in south and east Wales and in central England. The blueschists are interpreted as a sliver of exotic material caught within early transcurrent movements along the Menai Strait fault system. The Coedana Complex includes the *c*. 600 Ma Coedana Granite and a series of undated gneisses, and may have formed an older basement to the Monian Supergroup. However, because all boundaries between these two units are now tectonic, they are still to be treated as suspect unless new data can be produced to support a cover-basement relationship.

The suspect terrane model emphasises the importance of fault rocks in the Monian area and places special significance on the abrupt contrast in geology across the Menai Strait fault system. Such terrane analysis attempts to provide an essentially simple and fresh approach to the complexities and historical obfuscations that have arisen from conflicting interpretations of Monian geology. Because only tectonic boundaries exist between the four major units, the problem becomes one of examining these boundaries in some detail and of obtaining new data to test whether geological connections can be proven across the boundaries. Current programmes of geochemical, especially isotopic, analysis are likely to prove of fundamental value in this respect.

Since Greenly's detailed survey, progress in our understanding of Monian geology has proceeded erratically, with particularly major advances being achieved by the introduction and use of way-up criteria (Shackleton, 1954) and the recognition and interpretation of fault rocks (Baker, 1969, and others). Other apparent advances, such as the discovery of supposedly Cambrian, and even Ordovician, microfossils in the Monian Supergroup, have not yet stood up so well to the test of time (see references in Gibbons, 1983b and compare with Peat, 1984). The following sections initially summarise the main characteristics of the Monian Supergroup, the Coedana Complex, the Anglesey blueschists, and the Sarn Complex. The importance of overstep sequences across the Monian area is then emphasised, particularly with respect to constraints on the youngest age of Monian rocks and terrane docking times. Finally, a wider consideration is given to the position of the Monian area within the Caledonian-Appalachian Orogen.

3.2   The Monian Supergroup

Rocks belonging to this lithostratigraphic unit crop out extensively in northern and western Anglesey (Figure 3.2). Following Shackleton's revision of Greenly's original succession, the Supergroup sequence was recognised as consisting of four groups overlain by an acid volcanic formation (Shackleton, 1975). The lowest unit, known as the South Stack Group, consists of > 1000 m of well-bedded but polydeformed quartzites, turbiditic psammites

Figure 3.4   Southeast verging folds in South Stack Group psammite and pelites in the core of the Rhoscolyn anticline, Holy Island (see Shackleton, 1969; Lisle, 1987). The deformation seen in the pelite bed (lower left) is comparable to that seen within pelitic rocks in the overlying New Harbour Group (Figure 3.5).

Figure 3.5   Southeast verging folds within New Harbour Group pelites and thin psammites on the northwest limb of the Rhoscolyn Anticline, Holy Island.

and pelites and is spectacularly exposed in cliff sections along the southwest coast of Holy Island (Figures 3.4 and 3.5). Above the South Stack Group is a thick sequence (> 2000 m) of relatively monotonous and highly deformed pelites and semi-pelites known as the New Harbour Group. This group includes within it at least two horizons of metabasaltic lavas, and a horizon of serpentinised ultramafic rocks and gabbros.

There has been considerable debate on how one might relate the deformation histories of these two groups. Barber and Max (1979) have argued that because the New Harbour rocks are generally more deformed than those of the South Stack there is a tectonic break between the two groups. This deduction was incorporated into a model in which the New Harbour Group was deformed in a subduction zone and then thrust over the New Harbour Group. Several other authors, however, have emphasised the similarity between the polyphase deformation history of pelite beds within the South Stack Group and the more general deformation state typical of the New Harbour Group in western Anglesey (Gibbons, 1979; Maltman, 1979; Powell, 1979; Wood, 1979). Folded foliations and lineations are very common in the South Stack pelitic horizons, and it is clear that the deformation state is strongly influenced by lithological variations (Figures 3.4 and 3.5) (Cosgrove, 1980; Lisle, 1988).

Above the New Harbour Group Shackleton (1975) defined a dominantly volcaniclastic unit referred to as the Skerries Group (including the 'Church Bay Tuffs' and 'Skerries Grits' of Greenly, 1919). This group according to the Shackleton lithostratigraphic column, is overlain by the Gwna Group. The most characteristic feature of the Gwna lithologies is that the rocks are in the state of melange and therefore no internally coherent stratigraphy for this sequence is known. Greenly (1919) who introduced the term melange into the literature, envisaged a tectonic origin for the Gwna rocks, whereas Shackleton (1954) later reinterpreted the chaotic sequence as an olistostrome. A 'ghost' stratigraphy occurs within the Gwna melange on the Llŷn peninsula, where the structurally lowest part of the melange is rich in greywacke sandstone/pelite clasts, above which is a unit dominated by tholeiitic pillow lavas, followed by a mixed facies containing limestones, orthoquartzites, and many other sedimentary lithologies (Gibbons, 1980). Although this melange stratigraphy reveals nothing regarding the relative ages of the different melange clasts, it does suggest that the initial disruption that resulted in melange formation involved mostly intraformational deeper water sediments, followed by the later introduction of relatively 'exotic' lithologies of shallower water origin.

Shackleton (1975) placed Greenly's 'Fydlyn Group' above the Gwna melange. The stratigraphic status of both the Skerries and, particularly, the Fydlyn rocks has never been well defined. Both units have been awarded group status on Anglesey (Shackleton, 1969), although the Fydlyn rocks were later relegated to formational status (the Fydlyn Felsitic Formation of Shackleton, 1975). No way-up evidence has been described from the Skerries/ Gwna and Gwna/Fydlyn boundaries. Furthermore, clasts of New Harbour, Skerries, and Fydlyn lithologies may all be identified within the Gwna melange (Gibbons and Ball, in prep.). On the Llŷn peninsula, rocks supposedly lying above the melange and correlated with the Fydlyn lithologies on Anglesey (the Gwyddel Beds of Matley, 1928) have been shown to lie as clasts in the melange (Gibbons, 1980). A recent discovery using borehole data in northern Anglesey has shown the Gwna melange resting directly, with no structural break, upon New Harbour Group rocks (Gibbons and Ball, in prep.). In Northern Anglesey the rocks are at their lowest (anchizonal) metamorphic grade and the degree of overprinting deformation is also at a minimum. The New Harbour Group in this area is highly disrupted into crudely foliated and veined breccias by shearing subparallel to bedding, interpreted as having been induced when the rocks were still only partially lithified. This disruption is present to at least a depth of 200 m beneath the Gwna contact, and increases in intensity upwards to the Gwna contact, immediately beneath which the texture may be described as a New Harbour intraformational ('broken beds') melange.

The simplest interpretation of Monian Supergroup lithostratigraphy is therefore in terms of three major groups: a lowest, dominantly psammitic South Stack Group overlain by a dominantly pelitic New Harbour Group which, in turn is overlain by a Gwna melange containing exotic clasts including those of shallow water and volcanic origin. Neither stratigraphic base nor original top of this supergroup are exposed in northwest Wales. In northern and western Anglesey deformation and metamorphism each increase downwards through this sedimentary sequence, although both are enhanced within the more pelitic horizons, especially towards the base of the New Harbour Group. Two key unanswered questions relating to the supergroup are (1) how old is the sequence, and (2) when was it deformed? Foliated clasts of New Harbour Group material in an overlying Ordovician overstep sequence prove at least that the early fabric within the supergroup is pre-Fennian (Late Arenig). The prominent southeasterly verging folds so well developed in Holy Island (Lisle, 1988) could also be pre-Fennian but the possibility of a much younger age cannot yet be ruled out. The Gwna melange of Llŷn is involved in shear zones that are overlain unconformably by sediments no younger than Moridunian (lower Arenig) age. No radiometric data are yet available on the supergroup and palaeontological evidence remains equivocal but suggestive of a late Precambrian or early Cambrian age (Muir et al., 1979; Peat, 1984).

3.3   The Coedana Complex

This unit runs northeast–southwest for 27 km across the centre of Anglesey and is bounded on both sides by a tectonic contact with Monian Supergroup rocks (Figure 3.1). The Coedana Complex may be subdivided into high grade gneisses, the Coedana Granite, and fine grained rocks that have been interpreted as hornfelses to the granite. The same belt continues into northeast Anglesey as the Nebo Inlier (Greenly, 1919). Another small inlier of gneisses occurs in northwest Anglesey (the Gader inlier of Greenly, 1919) and may well belong to the same terrane.

The central Anglesey gneisses provide the only exposures of high grade Precambrian metamorphic rocks in England and Wales. The main outcrop of gneisses in Central Anglesey is dominated by basic amphibolites and variably migmatised almandine + sillimanite pelites. Some of the best examples of migmatitic and sillimanite-bearing pelites (Figure 3.6) occur in the Nebo Inlier (Greenly, 1919), within which also crop out calc-silicate lithologies containing ferrosalitic clinopyroxenes and grossular-rich garnets (Gibbons and Horák, 1990).

The Coedana Granite has provided a number of radiometric age determinations and appears to be c. 600 Ma old (see review of Monian radiometric age determinations in Gibbons and Horák, 1990). Greenly (1919) recognised megacrystic, pegmatitic, and muscovite-rich exceptions to a 'normal' granite facies, but no published work has elaborated on this initial description. A similar lack of modern detailed work has been published on the field relationships between this granite, the gneisses, and the rocks mapped by Greenly as hornfels. One recent advance in our understanding of the Coedana Complex, however, has come from the realisation that the granite is bounded to the southeast by a thick mylonite zone. These finely schistose mylonites, placed

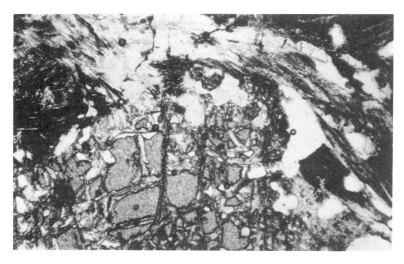

Figure 3.6   Almandine + sillimanite (top) + biotite in pelite from Coedana Complex. Field of view = 4 mm.

by Greenly (1919) within one of his 'Penmynydd Zones', mostly dip gently southeast away from the underlying granite and display a poorly developed ESE–WNW trending mineral lineation. Further thin section investigation of granitic rocks in central Anglesey beneath this Central Anglesey Shear Zone often reveals abundant evidence for both brittle and ductile responses to shearing deformation.

The northwestern margin of the Coedana Complex is only exposed at Llŷn Traffwl (GR SH 321 770) where granitic and gneissic rocks are in brittle fault contact with strongly deformed New Harbour metasediments. The excellent cataclasites developed at this contact contrast with the ductile mylonites along the southeast contact. It is notable that the New Harbour Group rocks exposed immediately northwest of the boundary with the Coedana Complex are very strongly deformed, so much so that Barber and Max (1979) argued for the presence of a mylonite belt along this margin. The present brittle fault contact at Llŷn Traffwl is interpreted as a relatively late (post- (and also probably syn-) Ordovician) fault that may have reactivated an originally more ductile tectonic contact.

Whereas the *c.* 600 Ma age of the Coedana granite provides the only repeatedly reproduced radiometric age on Anglesey, the age of the gneisses is still unknown. The age and origin of the 'hornfelses' are equally obscure. None of the Ordovician sediments on Anglesey have suffered the kind of ductile mylonitisation seen in the Central Anglesey Shear Zone, early movements along which were presumably older than this Ordovician over-step. The Central Anglesey Shear Zone separates the Coedana Complex from overlying, highly deformed metasediments that bear lithological similarities to parts of the Monian Supergroup. A particular similarity between lime-stone, quartzite and graphitic mudstone clasts in the Gwna melange (the 'Triple Group' of Greenly, 1919) and more highly deformed but broadly equivalent lithologies in the Central Anglesey Shear Zone may suggest some link between the two. Furthermore, clasts of granite present within the Gwna melange in Northern Anglesey (at Wylfa Head GR SH 353 943) are petro-graphically similar to the muscovite-rich facies of the Coedana granite. Because of this apparent (but tentative) lithostratigraphic link, one might argue that the Coedana Complex represents the basement to a Monian Supergroup cover sequence. However any originally unconformable contact between these two terranes has subsequently suffered localised but extreme and low angle ductile deformation. The general outcrop pattern of central Anglesey is reminiscent of extensional metamorphic core complexes, with a high grade 'basement' inlier lying in the footwall of a major shear zone above which lies a lower grade metasedimentary 'cover'.

3.4   The southeast Anglesey blueschists

Blueschists crop out within a strip of poorly exposed ground inland on southeast Anglesey. Most exposures show ophiolitic metabasites and mono-tonous semipelitic metasediments, with locally garnetiferous lithologies being spessartine-rich and interpreted as metamorphosed manganiferous sediment (Gibbons and Horák, 1990). Deformation is generally intense and pervasive (Figure 3.7a and b) although original micro-gabbroic igneous textures have been preserved within some of the larger metabasite masses. A blueschist overprint to a greenschist precursor has been recognised within the meta-basite lithologies (Gibbons and Gyopari, 1986) and actinolitic and sodic amphiboles from these two metamorphic events have yielded $^{40}Ar/^{39}Ar$ ages of c. 580 Ma and c. 550 Ma respectively (Dallmeyer and Gibbons, 1987). The blueschists in this area are extremely fresh and although most metabasite assemblages are dominated by epidote, exposures of lawsonite-bearing rocks are known from several localities along the east of the belt (Gibbons and Mann, 1983).

All contacts between these Monian blueschists and adjacent rocks are tectonic. The eastern margin of the belt is a zone of highly retrogressed semipelites and metabasites that appear to have been thrust over Gwna melange (Gyopari, 1984; the Bryn Menrig Shear Zone of Gibbons and Horák, 1990). Within the footwall of this shear zone Gwna lithologies show an unusually strong, N–S lineation. In contrast to what is at least in part a low-angle contact on the eastern margin of the belt, the northwest side of the schist belt is well exposed as a steep transcurrent shear zone (the Berw Shear Zone of Gibbons, 1987a). Within the Berw Shear Zone are slivers of recrystallised basic amphibolites and granitic rocks that have yielded Rb–Sr WR radio-

Figure 3.7 **a.** Pale phengitic metasedimentary schist (left) sub-isoclinally folded with metabasic blueschist in road cutting near Siglan Farm, southeast Anglesey. **b.** Close-up of isoclinal (sheath?) fold hinge in banded blueschist at same locality (field of view = 20 cm). The intense foliation exhibited by these rocks is accompanied by a N–S synmetamorphic mineral lineation.

metric ages of *c*. 590–560 Ma (the Holland Arms Gneisses of Beckinsale and Thorpe, 1979). It seems likely that these lithologies are slices of the Central Anglesey gneisses caught up within the shear zone.

Along strike to the southwest the Anglesey blueschist belt appears to die out, with only localised remnants of metabasic blueschists known from within a steep shear zone on the coast of Llŷn at Penrhyn Nefyn (Gibbons, 1981). This Llŷn Shear Zone may be a continuation of the Berw Shear Zone, with the same Gwna melange occurring along its western margin. However, a major difference on Llŷn is that the southeastern margin of the shear zone is occupied by the Sarn Complex.

3.5   The Sarn Complex

This is the most poorly exposed of the pre-Arenig terranes in northwest Wales. The most widespread lithology is that of the Sarn Granite, originally interpreted as intrusive into Arenig sediments but subsequently proven to underlie these sediments unconformably (Matley and Smith, 1936). Other lithologies include pyroxene-bearing diorites and apparently hybrid grano-dioritic rocks with blotchy xenolithic textures interpreted as due to magma-mixing between the granite and diorite (Gibbons, 1980; Beckinsale *et al.*, 1984). At Penrhyn Nefyn the Sarn Complex is represented by a tonalite.

In addition to these obviously igneous rocks, foliated lithologies described as gneisses also occur within the outcrop of the Sarn Complex. Some of these occur as foliated xenoliths within the plutonic rocks (as at Crugau Bach, GR 2125 2995), although in other cases the relationship is more obscure with apparent gradations existing between foliated and unfoliated plutonic rocks. Heavily retrogressed garnetiferous rocks are exposed at Parwyd on the southwest coast of Llŷn. These Parwyd gneisses (Figure 3.1) are isolated from all other rocks by faults, shear zones and a basal Arenig unconformity so that their relationship with the main outcrop of Sarn Complex rocks is unknown. However, further northeast within the main Sarn Complex outcrop, Matley described now unexposed garnetiferous gneisses (Matley, 1928; Gibbons, 1980) thus suggesting that the Parwyd gneisses may indeed be part of the Sarn Complex.

Rb–Sr whole-rock data on the igneous rocks of the Sarn Complex and on the Parwyd Gneisses have yielded poorly constrained 'errorchrons' inter-preted as defining a *c.* 550–540 Ma age for the complex. Further isotopic information is presently being collected in order to enhance our understanding of the Rb–Sr data (J. M. Horák, pers. commun., 1989). Finally, highly cataclastic granitic and dioritic rocks occur as clasts within the Gwna Melange on Bardsey Island off the extreme southwestern tip of the Llŷn peninsula (Matley, 1913; Gibbons, 1980). Such clasts are extremely rare within the Gwna melange, the only other examples so far recorded being those at Wylfa Head. Present work is investigating the possibility of a connection between the Bardsey Granites and either the Sarn Complex or the granitic rocks in Central Anglesey (J. M. Horák, pers. commun., 1990).

3.6   Regional correlations

Rocks of possible Monian affinity have been identified in southeast Ireland,

Newfoundland, and in several areas along the eastern seaboard of mainland North America (see reference list in Gibbons, 1983b). The most convincing of these correlations is the direct comparison between the South Stack Group and similar metasediments in southeast Ireland known as the Cullenstown Formation (Crimes and Dhonau, 1967). More recently Tietzsch-Tyler and Phillips (1989) have included the Cullenstown Formation within a larger Cahore Group, and attempted a more detailed and specific correlation between formations within the Cahore and South Stack Groups. One important corollary of such a correlation is that if the two groups are age-equivalents, then the South Stack Group is indeed Cambrian in age (*cf.* Barber and Max, 1979; Muir *et al.*, 1979).

The Cullenstown Formation is bounded to the southeast by a wide, sinistrally transcurrent mylonite zone of pre-Arenig age (Murphy, 1990). Within this subvertical shear zone Cullenstown Formation metasediments are strongly mylonitised against sheared gneissic and plutonic rocks of the Rosslare Complex. Gabbroic rocks within the Rosslare Complex have yielded an $^{40}Ar/^{39}Ar$ age of 618 ± 5 Ma (Max and Roddick, 1989), with slightly older ages being obtained from surrounding gneisses. It is possible that the Rosslare Complex is equivalent to the Central Anglesey Gneisses, in which case the Llŷn Traffwl fault zone on Anglesey may be interpreted as the brittle, reactivated equivalent of the Cullenstown/Rosslare shear zone contact. However, the presence of a wide vertical zone of sinistral mylonites in southeast Ireland is comparable to exposures in the Berw shear zone in southeast Anglesey, so that the Rosslare Complex may lie within a continuation of the Menai Strait line.

If the Cullenstown Formation is part of the Cahore Group then a Cullenstown–Monian correlation links the Monian Supergroup with the Lower Palaeozoic geology of southeast Ireland. By contrast, no such connection may be made between Monian rocks and late Precambrian–early Cambrian rocks exposed to the southeast of the Menai Strait fault system. Instead, the Sarn Complex has much closer lithological affinity with the mostly igneous and calc-alkaline late Precambrian basement exposed in central England and around the southern and eastern periphery of the Welsh Basin. The Menai Strait line is therefore taken to define the northwestern boundary of the Avalonian Superterrane—a group of mostly late Precambrian to early Cambrian, arc-related terranes dispersed along the margin of Gondwanaland.

3.7   Overstep sequences and age constraints

The identification across a suspect terrane boundary of overstep sequences of known age is crucial to elucidating the docking history of terranes along that boundary. Such overstep sequences are particularly important in northwest

Figure 3.8   The Arenig overstep sequence at Ogof Gynfor on the north coast of Anglesey rests unconformably upon quartzite clasts within the Gwna melange (lower left). Detrital fragments of the melange have been incorporated into the sediment (top right).

Wales where age control on Monian rocks and their deformation is generally very poor. Both pre- and post-Arenig deformation events are known to have affected the Anglesey/Llŷn area, yet it is often difficult to assign a fold phase at a given locality to an event of known age. The most obvious overstep sequence across northwest Wales is that produced by Arenig marine transgression, the sediments of which were deposited with marked unconformity upon melange, gneisses, Coedana Granite and mylonites of the Llŷn shear zone (Figure 3.8). Palaeontological evidence indicates that this transgression took place no later than during Moridunian (lower Arenig) times on Llŷn and Fennian (upper Arenig) times on Anglesey (Beckly, 1987).

In addition to the Arenig overstep, four outliers of sediments and volcanic rocks rest upon Monian rocks in Anglesey. These four outliers comprise the Trefdraeth Conglomerate (on the Gwna melange in central Anglesey), the Careg Onen Beds (on the blueschist belt and Gwna melange in northeast Anglesey), the Baron Hill Beds (on Gwna melange in eastern Anglesey), and the Bwlch Gwyn Tuff (on basic amphibolitic mylonites in the Berw shear zone). Greenly (1919) identified Monian fragments in the often volcaniclastic sediments of these outliers and interpreted them as resting unconformably upon a Monian basement. Furthermore, the outliers have been correlated with the Arfon Group, a sequence of rhyolitic ash-flow tuffs and mostly volcaniclastic sediments that crop out on the Welsh mainland just southeast of the Menai Strait (Greenly, 1919; Reedman et al., 1984). The Arfon Group lies conformably beneath fossiliferous sediments of at least late Lower Cambrian age (Howell and Stubblefield, 1950).

The supposedly 'Arfon Group' outliers on Anglesey are of critical importance because they provide the only potentially unequivocal evidence to prove that Monian rocks had been formed, deformed and moved to their present position by at least late Lower Cambrian times. Without them, only the Arenig overstep places a minimum age on the deformation and docking of Monian terranes. Fragments of Monian-type lithologies are present in Lower and Middle Cambrian rocks in northwest Wales (Greenly, 1919, 1923; Woodland, 1938; Gibbons, 1983b) and suggest that Monian rocks were available for erosion by this time. An interesting discovery of blue amphibole detritus in early or Middle Cambrian sediments in southwest Wales (St. Non's Sandstone; Mackay, 1987) again suggests the early exposure of a Monian detrital source, but none of this included fragment evidence provides as compelling an argument as would an unconformity of proven Arfon Group upon Monian basement. The case for Arfon Group outliers on Anglesey is strong but not unequivocal and further detailed studies are in progress (R. Brewer, pers. commun., 1989).

The importance of the 'Arfon Group' outliers on Anglesey has been heightened by the interpretation of a Lower Palaeozoic age for the Monian Supergroup (Barber and Max, 1979; Teitzsch-Tyler and Phillips, 1989). Barber and Max (1979) argued that because there is little apparent difference in metamorphic grade and structural complexity between the Monian Supergroup and overlying Arenig sediments in Northern Anglesey, no great time break exists between these two units in this area. The proposal was supported by micropalaeontological evidence for a Lower Cambrian age for the Gwna melange (Muir et al., 1979) although this has subsequently been criticised (Peat, 1984) and cannot now be taken as well established. If one or more of the three outliers in eastern Anglesey are truly part of the Arfon Group sequence, and if they rest unconformably upon Monian schists and Gwna melange, then the latter must be no younger than late Lower Cambrian in age. Similarly, if the Trefdraeth Conglomerate can be shown to belong to the Arfon Group, then the same age constraint applies to the Gwna melange of central Anglesey.

Even if it can be demonstrated that the melanges of eastern and central Anglesey are of pre–late Lower Cambrian age, then in order to apply this to western and northern Anglesey one must assume that all the Gwna melange outcrops are of the same age. Some previous workers (Barber and Max, 1979; Wood, 1979) have argued that Gwna melanges occur at more than one stratigraphic horizon, although no detailed field evidence has been presented to support this. Detailed work on the Gwna melange exposed on Llŷn has revealed no evidence for any stratigraphic break separating different melanges, and it is notable that all of the many distinctive lithological facies seen within the Llŷn melange (Gibbons, 1980) can be identified in exposures across Anglesey. Despite being extremely heterogeneous, the Gwna melange is lithologically distinctive and similar in all outcrops mapped by Greenly

(1919) and Matley (1928): a strong, detailed argument would be required to prove that there are Gwna melanges of different ages.

In summary, there is no doubt that all Monian rocks are pre-Fennian in age. This conclusion also applies to the high grade gneissic metamorphism in central Anglesey, the blueschist metamorphism in southeast Anglesey (dated as *c.* 550 Ma), the transcurrent shearing that produced the Berw shear zone, and at least the earliest deformation phase in the Monian Supergroup. A slightly older age constraint is provided by the Moridunian (or older) unconformity on the mylonites of the Llŷn shear zone and the Parwyd gneisses on the Llŷn peninsula. The presence of Monian-type fragments in late Lower Cambrian and Middle Cambrian sediments on the Welsh Mainland, and the possible presence of Arfon Group outliers on Anglesey, all point to the exposure and erosion of Monian rocks by the late Lower Cambrian. The latter conclusion, however, apparently conflicts with a correlation between the Monian Supergroup and the Cahore Group in southeast Ireland: such a correlation therefore needs to be examined critically before it is used to overrule the included fragment/Arfon Group evidence. The polyphase deformation of the Monian Supergroup presumably includes some component of Acadian deformation (as the Ordovician on Anglesey and Llŷn is deformed), but it remains unclear whether the obvious southeastward verging folds so well exposed on Holy Island and elsewhere are pre- or post-Fennian. The fact that these folds are associated with a well-developed micaceous foliation of a distinctly higher grade than anything seen in the Ordovician succession might suggest that they are older: a radiometric age on this fabric is required.

3.8 Monian tectonics: an overview

Monian rocks are uniquely important in recording late Precambrian and probably early Cambrian events along the Avalonian/Cadomian margin of Gondwanaland (Figure 3.9). As Dewey (1969), Wood (1974) and others have emphasised, the presence of blueschists suggests palaeosubduction. However, the recognition of transcurrent terrane boundaries in this area indicates that any former subduction zone has been dismembered, presumably by oblique plate motions. Monian rocks may therefore be interpreted as fault-bounded slivers accreted to the margin of an Avalonian Superterrane, much of which was itself produced by subduction-related magmatic, metamorphic and sedimentary processes. The amount of lateral transport between Monian and Avalonian terranes is completely undetermined, although one may speculate upon the likely tectonic settings appropriate for each unit:

Blueschists: subducted oceanic crust and sedimentary cover. Subduction took place at around 550 Ma, possibly preceded by sub-ocean floor meta-

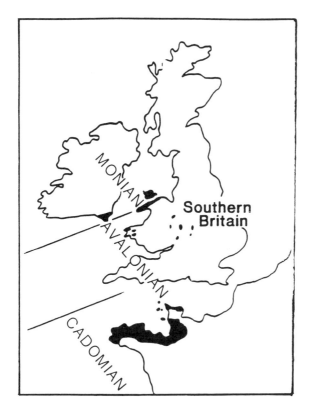

Figure 3.9   Sketch map of the British Isles and northwest France illustrating the relative position of Monian rocks in northwest Wales and southeast Ireland, Avalonian rocks in southwest Wales and central England, and Cadomian rocks in the Channel Islands, Normandy and Brittany. The boundary between Monian and Avalonian rocks is interpreted to lie along the Menai Strait Fault System.

morphism at *c.* 590 Ma, and followed by rapid uplift and tectonic transport against the Gwna melange prior to the eruption of the Arfon Group in late Lower Cambrian times (if Arfon Group correlations hold true). The blueschist metamorphism may actually be recording the dynamic obduction of these rocks rather than their initial burial, as seems often to be the case for blueschists preserved on continental crust (e.g. Gibbons and Horák, 1984; Gibbons, 1987b).

Monian Supergroup: Whereas the South Stack Group is dominated by mature quartzites and turbidites that suggest a stable cratonic provenance, the New Harbour Group shows a strong volcanic arc influence, both in its sedimentary and igneous petrology. The tectonic setting therefore appears to have been marginal to an arc built on continental crust, perhaps in a forearc setting with the incipient production of oceanic crust by extension above the subducting slab (*cf.* the present day Andaman Sea). Sedimentation in this marine basin was terminated by chaotic disruption of the upper part of the New Harbour Group and the introduction of exotic clasts, including much shallower water lithologies, to form the Gwna melange. A collisional accretion process is suggested as an explanation for the generation of this melange, perhaps similar to the Lichi melange produced in Taiwan by the collision of an arc system with a continental margin (Page and Suppe, 1981).

Monian gneisses and granites: These might represent the provenance material for the South Stack Group, and/or the continental unit colliding with the Avalonian margin. It is possible that this terrane represents the roots of an arc built on continental crust. Further isotopic and whole-rock geochemical work in progress may provide more control on the age of the gneisses and the likely tectonic setting of this terrane.

Sarn Complex: a calc-alkaline plutonic complex interpreted as part of the Avalonian arc against which the Monian rocks were accreted.

The transcurrent dispersal of Cadomian, Avalonian and Monian terranes along the Gondwanan margin was effected along major fault systems that typically provided a focus for later Phanerozoic (usually dip-slip) reactivation. It is likely that prominent lineaments in Britain, such as the Welsh Borderlands and the Malverns Fault System, formed part of a larger family of fractures linking in to the Menai Strait line. Similarly, old, steep mylonite belts are known from the Cadomian exposures in northern Brittany (Strachan *et al.*, 1989) and the Avalonian of Eastern Canada (Gibbons, 1990); these testify to the existence of an anastomosing network of transcurrent faults dispersing terranes along the Gondwanan margin in late Precambrian–early Cambrian times. Subsequent phases of accretion later during the Palaeozoic, terminated by final oceanic closure, constructed the collage of terranes that make up the present outcrop pattern of the Caledonian-Appalachian orogen.

Acknowledgements

Assistance with the preparation of Figures 3.1, 3.2 and 3.6 was provided by J. Horák. Figure 3.3 was drawn by Emrys Phillips.

References

Baker, J. W. (1969) Correlation problems of unmetamorphosed Precambrian rocks in Wales and southern Ireland. *Geol. Mag.* **106**, 246–259.
Barber, A. J. and Max, M. D. (1979) A new look at the Mona Complex (Anglesey, North Wales). *J. Geol. Soc. London* **136**, 407–432.
Bassett, M. G., Bevins, R. E., Gibbons, W. and Howells, M. F. (1986) A geotraverse through the

Caledonides of Wales. In *Synthesis of the Caledonian Rocks of Britain*, eds. Fettes, D. J. and Harris, A. L., *NATO ASI Ser. C, mathematical and physical science* **175**, 29–76.

Beckinsale, R. D. and Thorpe, R. S. (1979) Rubidium–strontium whole-rock evidence for the age of metamorphism and magmatism in the Mona Complex of Anglesey. *J. Geol. Soc. London* **136**, 433–439.

Beckinsale, R. D., Evans, J. A., Thorpe, R. S., Gibbons, W. and Harmon, R. S. (1984) Rb–Sr whole-rock isochron ages, $\delta^{18}O$ values and geochemical data for the Sarn Igneous Complex and the Parwyd gneisses of the Mona Complex of Llyn, N. Wales. *J. Geol. Soc. London* **141**, 701–709.

Beckly, A. J. (1987) Basin development in North Wales during the Arenig. *Geol. J.* **22**, 19–30.

Blake, J. F. (1888) On the Monian System of rocks. *Q. J. Geol. Soc. London* **44**, 463–546.

Cosgrove, J. W. (1980) The tectonic implications of some small scale structures on the Mona Complex, Holy Isle; North Wales. *J. Struct. Geol.* **2**, 383–396.

Crimes, T. P. and Dhonau, N. B. (1967) The Precambrian and Lower Palaeozoic rocks of Southeast County Wexford, Eire. *Geol. Mag.* **104**, 213–221 and 400.

Dallmeyer, R. D. and Gibbons, W. (1987) The age of blueschist metamorphism in Anglesey, north Wales: evidence from $^{40}Ar/^{39}Ar$ mineral dates of the Penmynydd Schists. *J. Geol. Soc. London* **144**, 843–850.

Dewey, J. F. (1969) Evolution of the Appalachian-Caledonian orogen. *Nature (London)* **222**, 124–148.

Gibbons, W. (1979) Discussion on Barber, A. J. and Max, M. D., 1979. A new look at the Mona Complex. *J. Geol. Soc. London* **136**, 429–430.

Gibbons, W. (1980) The geology of the Mona Complex of the Lleyn peninsula and Bardsey Island: North Wales. Unpublished Ph.D. thesis, Portsmouth Polytechnic.

Gibbons, W. (1981) Glaucophanic amphibole in a Monian Shear Zone in Llyn, North Wales. *J. Geol. Soc. London* **138**, 139–143.

Gibbons, W. (1983a) The Monian Penmynydd Zone of Metamorphism in Llyn, North Wales. *Geol. J.* **18**, 1–21.

Gibbons, W. (1983b) Stratigraphy, subduction and strike-slip faulting in the Mona Complex of North Wales—a review. *Proc. Geol. Assoc.* **94**, 147–163.

Gibbons, W. (1987a) The Menai Strait Fault System: an early Caledonian terrane boundary in North Wales. *Geology* **15**, 744–747.

Gibbons, W. (1987b) Discussion on Chopin, C., 1987. Very high-pressure metamorphism in the Western Alps. *Philos. Trans. R. Soc. London* **321**, 195.

Gibbons, W. (1989) Suspect terrane definition in Anglesey, North Wales. *Geol. Soc. Am. Spec. Pap.* **230**. Terranes in the Circum-Atlantic Orogens, Dallmeyer, R. D. (ed.), 59–66.

Gibbons, W. (1990) Trancurrent ductile shear zones and the dispersal of the Avalon Super-terrane. In *The Cadomian Orogeny*, eds. D'Lemos, R. S., Strachan, R. A. and Topley, C. G., *Spec. Publ. Geol. Soc. London* **51**, 407–423.

Gibbons, W. and Ball, M. J. (1989) A revision of Monian Supergroup stratigraphy on Anglesey. In preparation.

Gibbons, W. and Gyopari, M. C. (1986) A greenschist protolith for blueschists in Anglesey, U.K. In *Blueschist and Eclogites*, eds. Evans, B. W. and Brown, E. H., *Geol. Soc. Am. Mem.* **164**, 217–228.

Gibbons, W. and Horák, J. M. (1984) Alpine metamorphism of Hercynian hornblende granodiorite beneath the blueschist facies *schistes lustrés* nappe of NE Corsica. *J. Metamorph. Geol.* **2**, 95–113.

Gibbons, W. and Horák, J. M. (1990) Contrasting metamorphic terranes in Northwest Wales. In *The Cadomian Orogeny*, eds. D'Lemos, R. S., Strachan, R. A. and Topley, C. G., *Spec. Publ. Geol. Soc. London* **51**, 315–327.

Gibbons, W. and Mann, A. (1983) Pre-Mesozoic lawsonite in Anglesey, North Wales—the preservation of ancient blueschists. *Geology* **11**, 3–6.

Greenly, E. (1919) *The Geology of Anglesey*. 2 vols., Mem. Geol. Surv. U.K., London.

Greenly, E. (1923) The succession and metamorphism in the Mona Complex. *Q. J. Geol. Soc. London* **79**, 334–351.

Gyopari, M. C. (1984) A study of blueschist mineral chemistry and a new look at the Penmynydd–Gwna boundary in SE Anglesey. Unpublished M.Sc. thesis, University of Wales, Cardiff.

Howell, B. F. and Stubblefield, C. J. (1950) A revision of the fauna of the North Welsh *Conocorphyte viola* beds implying a Lower Cambrian age. *Geol. Mag.* **87**, 1–16.

Lisle, R. J. (1988) Anomalous vergence patterns on the Rhoscolyn Anticline: implications for structural analysis of refolded regions. *Geol. J.* **23**, 211–220.

Mackay, A. W. (1987) The sedimentology and petrology of the Cambrian sediments below the *Lingula* Flags on the St. David's Peninsula, South West Wales. Unpublished M.Sc. thesis, University of Wales, Cardiff.

Maltman, A. J. (1979) Discussion of Barber, A. J. and Max, M. D., 1979. A new look at the Mona Complex. *J. Geol. Soc. London* **136**, 428.

Mann, A. (1986) Geological studies within the Mona Complex of Central Anglesey, North Wales. Unpublished Ph.D. thesis, University of Wales, Cardiff.

Matley, C. A. (1913) The geology of Bardsey Island. *Q. J. Geol. Soc.* **69**, 514–533.

Matley, C. A. (1928) The Precambrian Complex and associated rocks of S.W. Lleyn (Carnarvonshire). *Q. J. Geol. Soc. London* **84**, 440–504.

Matley, C. A. and Smith, B. (1936) The age of the Sarn Granite. *Q. J. Geol. Soc.* **92**, 188–200.

Max, M. D. and Roddick, J. C. (1989) Age of metamorphism in the Rosslare Complex, SE Ireland. *Proc. Geol. Assoc.* **100**, 113–123.

Muir, M. D., Bliss, G. M., Grant, P. R. and Fischer, M. (1979) Palaeontological evidence for the age of some supposedly Precambrian rocks in Anglesey, North Wales. *J. Geol. Soc. London* **136**, 61–64.

Murphy, F. C. (1990) Basement-cover relationships of a reactivated Cadomian mylonite zone: Rosslare Complex, SE Ireland. In *The Cadomian Orogeny*, eds. D'Lemos, R. S., Strachan, R. A. and Topley, C. G., *Spec. Publ. Geol. Soc. London* **51**, 329–339.

Page, B. M. and Suppe, J. (1981) The Pliocene Lichi melange of Taiwan: it's plate tectonic and olistostromal origin. *Am. J. Sci.* **281**, 193–227.

Peat, C. J. (1984) Comments on some of Britain's oldest microfossils. *J. Micropalaeontol.* **3**, 65–71.

Powell, D. (1979) In discussion on Barber, A. J. and Max, M. D., 1979. A new look at the Mona Complex. *J. Geol. Soc. London* **136**, 427–428.

Reedman, A. J., Leveridge, B. E. and Evans, R. B. (1984) The Arvon Group ('Arvonian') of north Wales. *Proc. Geol. Assoc.* **95**, 313–322.

Shackleton, R. M. (1954) The structure and succession of Anglesey and the Lleyn Peninsula. *Adv. Sci.* **11**, 106–108.

Shackleton, R. M. (1969) The Precambrian of North Wales. In *The Precambrian and Lower Palaeozoic Rocks of Wales*, ed. Wood, A., University of Wales Press, Cardiff, 1–22.

Shackleton, R. M. (1975) Precambrian Rocks of Wales. In *A Correlation of the Precambrian Rocks in the British Isles*, eds. Harris, A. L. *et al.*, *Geol. Soc. London Spec. Rep.* No. **6**, 76–82.

Strachan, R. A., Treloar, P. J., Brown, M. J. and D'Lemos, R. S. (1989) Cadomian terrane tectonics and magmatism in the Armorican Massif. *J. Geol. Soc. London* **146**, 423–426.

Tietzsch-Tyler, D. and Phillips, E. (1989) Correlation of the Monian Supergroup in NW Anglesey with the Cahore Group in SE Ireland. *J. Geol. Soc. London* **146**, 417–418.

Wood, D. S. (1974) Ophiolite, mélanges, blueschists, and ignimbrites: early Caledonian subduction in Wales? In *Modern and Ancient Geosynclinal Sedimentation*, eds. Dott, R. H. and Shaver, R. H., *Soc. Econ. Paleontol. Mineral. Spec. Publ.* **19**, 187–194.

Wood, D. S. (1979) In discussion on Barber, A. J. and Max, M. D., 1979. A new look at the Mona Complex (Anglesey, North Wales). *J. Geol. Soc.* **136**, 425–427.

Woodland, A. W. (1938) Petrological studies in the Harlech Grit series of Merionithshire. *Geol. Mag.* **75**, 366–382, 440–454, 529–539.

4 The Rosslare Complex: a displaced terrane in southeast Ireland

J. A. WINCHESTER, M. D. MAX and F. C. MURPHY

4.1 Introduction

The Precambrian Rosslare Complex (Baker, 1970; Max and Dhonau, 1971; Max, 1975, 1978; Barber *et al.*, 1981; Winchester and Max, 1982; Max and Long, 1985) comprises a varied suite of gneisses exposed in the extreme southeast of Ireland and the adjacent offshore area (Figure 4.1). It crops out as a tabular body of gneisses and igneous rocks bounded by steep mylonite zones which may be linked with suspect terrane boundaries recorded to the northeast in Anglesey. It may either represent a detached slice of the Avalon Superterrane, which extends from New England and southeast Newfoundland to the English Midlands (Williams and Max, 1980), or it could represent a far-travelled basement terrane. Although long regarded as Precambrian basement (Calloway, 1881), the Complex has only recently yielded isotopic indications of its antiquity (Max and Roddick, 1989). Because all exposed contacts with adjacent rocks are tectonic, field evidence has never provided clear proof that the gneisses in the complex comprise basement to Lower Palaeozoic rocks in Leinster. Exposure of the Rosslare Complex is poor: the best coastal outcrops occur around St. Helens and Kilmore Quay, while inland there are only a few small scattered exposures (Figure 4.1).

A broad belt of mylonites bounds the Rosslare Complex to the northwest. These mylonites, termed by Murphy (1990) the Ballycogly shear zone, occupy a 4 km-wide area between the Rosslare Complex and a basin in which upper Palaeozoic and early Mesozoic rocks are preserved. Max and Dhonau (1971) considered that these mylonites were largely derived by shearing of the Complex itself. Baker (1970) divided these mylonites into a gneissic protolith (Tomhaggard Zone) cropping out south of a metasedimentary protolith (Ballycogly Group). Baker considered that the Ballycogly Group represents a highly deformed late Precambrian 'Monian' cover to the Rosslare Complex. Between these two zones there crops out an undated metasedimentary unit, the Silverspring Beds (Baker, 1966), which was thought to represent the deformed equivalents of the Arenig Tagoat Group occurring along strike to the east (see Figure 4.1). The oldest rocks northwest of the mylonites bounding the Rosslare Complex comprise the undated, but probable lowest

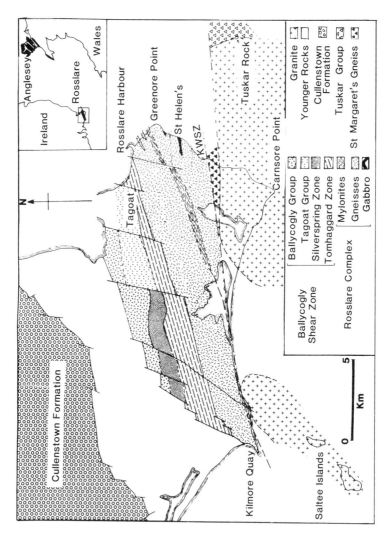

Figure 4.1 Location map of the Rosslare Complex showing the distribution of gneisses and adjacent younger rocks, including Palaeozoic Granites.

Cambrian Cullenstown Formation and Bray Group, which consist of shelf-derived clastic sediments and greywackes that are not seen to be cover to the Rosslare Complex (Max and Dhonau, 1974; Bruck et al., 1979).

To the southeast the Complex is bounded by a shear zone (Kilmore-Wilkeen Mylonite Zone) separating it from the St. Margaret's Gneisses (Baker, 1970) which are now thought to be a deformed portion of the Tuskar Group of possible Cambro-Ordovician age, exposed on the Tuskar Rock and adjacent submarine outcrops (Max and Ryan, 1986). The Rosslare Complex structural block may therefore be considered as a discrete terrane since it is bounded by major structural lineaments which separate it from adjacent, but apparently unrelated and younger tectonosedimentary blocks. Attempts to correlate between the Rosslare Complex and the superficially similar rocks of the Mona Complex of Anglesey have failed to establish any clear connection, although both complexes are exposed within a linear NE–SW trending basement high exposing Precambrian rocks. The long tectonometamorphic history of the Rosslare Complex, however, suggests that it may contain older gneisses than any recognized in the Mona Complex. These gneisses are probably the oldest rocks exposed in the British Isles southeast of the Iapetus Suture, and may be comparable in age to the Icartian Gneisses exposed in the Armorican Massif of Brittany and the Channel Isles.

4.2   The Rosslare Complex: constituents and geological history

Four tectonometamorphic events have been recognized in the Rosslare Complex (Max, 1975; Max and Long, 1985; Max and Roddick, 1989). Of these, the earlier two were Precambrian, subsequent episodes were Palaeozoic (Table 4.1).

4.2.1   Early tectono-thermal history

The age of the earliest metamorphism that appears to have affected both the dark and grey gneisses (Max, 1975) is poorly constrained. Nd model ages of 1.7 Ga (Davies et al., 1985) suggest the presence of an ancient component, but their significance is uncertain, especially as the samples dated and described by Davies et al. (1985) as granitic gneiss were collected from semipelitic grey gneisses of sedimentary origin at St. Helen's and Kilmore Quay. Poorly constrained Rb–Sr data (Max, 1975) from granodioritic and grey gneiss also suggests an age as old as 2.0 Ga. The earliest metamorphism therefore predates the date of 626 ± 6 Ma, (Max and Roddick, 1989), which represents the peak of the second event (Table 4.1). The rocks produced by the early metamorphism are grey biotite-garnet-quartz-feldspar gneisses and dark amphibolitic gneisses, referred to respectively as the Grey Gneiss and Dark Gneiss. The metamorphism was accompanied by the production of coarse-grained trondhjemitic migmatites and a gneissose texture throughout the Grey and Dark Gneiss.

Table 4.1   Chronological sequence of tectonic, intrusive and metamorphic events within the Rosslare Complex (Max and Long, 1985; Max and Roddick, 1989; Murphy, 1990)

	Deformation and metamorphism	Intrustions/volcanics	Stratigraphy	(Ma)
Post-Event 4	Mesozoic and Tertiary fault reactivation associated with development of nearby Celtic Sea basin			Post- 100—
				Silurian
	Carboniferous and Permo-Triassic fault reactivation in shear zones during sedimentation			200—
Event 4	Minor renewed shearing in shear zones and deformation of Saltees Granite. Low greenschist facies retrogression in Rosslare Complex	Carnsore Granite (432 ± 3 Ma) Younger Basic Dykes Saltees Granite (436 ± 7 Ma)		〰〰〰 Silurian
				450—
Event 3	Upper greenschist facies metamorphism within Rosslare Complex, with K–Na metasomatism (480 Ma) $S2_{myl}$ cleavage with dip-slip movement in Ballycogly Mylonite Zone. S1 cleavage in Silverspring, Tagoat and Tuskar Groups	Tuskar volcanics ↑	Tuskar Group Tagoat group Silverspring Beds	Ordovician 500—
				Cambrian
		Older Basic Dykes (age poorly constrained)		550—
			Cullenstown Group Ballycogly Group (precise age uncertain)	Precambrian 600—
Event 2	Amphibolite facies metamorphism within Rosslare Complex. Sinistral shearing and greenschist facies metamorphism in Bounding Mylonite Zones ($S1_{myl}$) [626 ± 6 Ma]	↓ St. Helen's Gabbro (618 ± 5 Ma) granodiorite sheets		〰〰〰 1000—
Event 1	High amphibolite facies metamorphism migmatization. development of gneisses	Greenore diorite Dark Gneiss	'Grey Gneiss' (deposited as greywacke) 1.7 Ga Sm/Nd age on Grey Gneiss	1500—

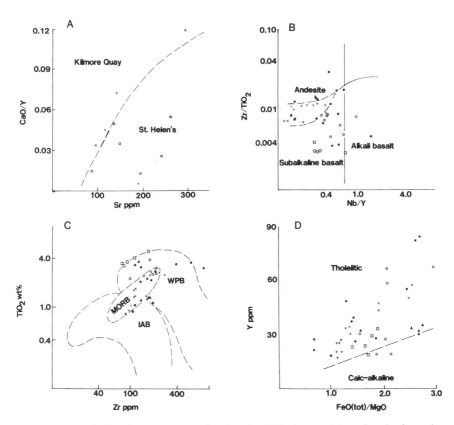

Figure 4.2 Variation diagrams contrasting the chemical characteristics of rocks from the Rosslare Complex. (A) A CaO/Y–Sr diagram discriminating between Grey Gneiss from St. Helen's (circles) and Kilmore Quay (crosses). (B) Zr/TiO_2–Nb/Y binary plot (Winchester and Floyd, 1977) used to illustrate that almost all the basic intrusive rocks have a subalkaline basaltic chemistry. Samples from the Greenore Diorite plot in the 'andesite' field. Symbols: Dark Gneiss—filled circles; St. Helen's Gabbro—open circles; Older Basic Dykes—upright crosses; Younger Basic Dykes—diagonal crosses; Greenore Diorite—triangles; Tuskar Metavolcanics—squares. (C) TiO_2–Zr binary diagram (Pearce, 1979) indicating that the Rosslare basic rocks spread within the MORB and WPB fields, a scatter often characteristic of continental tholeiitic rocks. (Symbols as in B.) (D) Y–FeO(total)/MgO variation diagram illustrating the tholeiitic nature of most basic suites. (Symbols as in B.)

4.2.1.1 Grey Gneiss Geochemical studies have suggested that the Grey Gneiss formerly consisted dominantly of greywackes containing detritus derived from an igneous provenance. Minor compositional differences which occur between the Grey Gneiss from Kilmore Quay and that from St. Helen's were revealed in an earlier geochemical study of the Rosslare Complex (Winchester and Max, 1982). In particular, the Grey Gneiss from Kilmore Quay was found to contain more CaO and less Ni and Y than the Grey Gneiss at St. Helen's. Since this original study additional trace element determinations have revealed that the Kilmore Quay Grey Gneiss contains lower Ce, La and Zr, and higher Cu than those at St. Helen's. However, the apparent difference in P_2O_5 results from the inclusion of a single relatively phosphatic sample also enriched in Y and LREE. These differences in composition are nonetheless relatively small, and only variation diagrams using CaO (Figure 4.2A) achieve a clear separation of the Grey Gneiss into two separate fields. As they are based on a small sample set, and a larger population could well show some overlap of fields, we presently consider that all the Grey Gneiss belongs to the same broad metasedimentary group.

4.2.1.2 Dark Gneiss   Often intimately intermixed with the Grey Gneiss is a dark, amphibolitic gneiss. Precise field relations between this 'Dark Gneiss' and the Grey Gneiss are obscured by intense deformation and meta-morphism, but geochemical studies show the Dark Gneiss to have a broadly basaltic composition (Winchester and Max, 1982) and, using a Zr/TiO_2-Nb/Y discriminant diagram (Winchester and Floyd, 1977; Floyd and Winchester, 1978), are shown to display a range of subalkaline basaltic compositions (Figure 4.2B). At Kilmore Quay an ultramafic pod was sampled which has a high Nb/Y ratio and appears to be compositionally unrelated to any other Dark Gneiss. Within the basaltic Dark Gneiss the range of compositions might result in part from selective element mobility during multiple metamorphic events, and for this reason most of the chemical studies concentrate on the use of the relatively immobile elements Ti, Zr, Y, Nb and LREE. However, we recognize that during prolonged high grade meta-morphism complete immobility of even these elements is unlikely. Therefore, if the compositional scatter shown by the Dark Gneiss is not purely a result of element mobilization during metamorphism, it suggests that the Dark Gneiss may comprise more than one basic intrusive or extrusive suite within the Grey Gneiss. Full REE determinations of Rosslare Complex rocks have not been completed, but if $(La/Y)_{cn}$ and $(Ce/Y)_{cn}$ ratios are used as indicators of LREE/HREE ratios, the Dark Gneiss can be regarded as generally enriched in LREE. This suggests that the rising basic magma was enriched in LILE during its ascent, indicating that it was probably intruded through continental crustal material before its final emplacement in the Grey Gneiss. Such a model implies that continental basement lay beneath the Rosslare Complex at that time.

On a TiO_2–Zr binary plot (Pearce, 1979) (Figure 4.2C) the Dark Gneiss falls within the WPB and MORB fields, consistent with the scatter exhibited by gabbro or dolerite suites of continental tholeiitic affinity. Concentrations of TiO_2, Zr, Y, P_2O_5, Nb, Cr, Ni and LREE are highly variable, suggesting that considerable fractionation has occurred. High concentrations of K_2O, Rb and Ba, atypical of fresh gabbro or basalt, suggest introduction of these elements during metamorphism. High Cl values (Table 4.1) probably result from recent seawater contamination.

4.2.1.3 Greenore Diorite   North of St. Helen's there are isolated beach exposures of a coarsely foliated epidote-sericite-quartz gneiss of intermediate composition with relict garnets of up to 12 mm across (Greenore Diorite) which was also affected by the earliest metamorphism (Winchester and Max, 1982). The presence of rare, pale green, probably metamorphic clinopyroxene intergrown with brown biotite in this rock, together with the occurrence of brown amphibole in the Dark Gneiss, indicates that at least high amphibolite facies metamorphism affected these rocks.

The Greenore Diorite is chemically unique within the Complex. Unlike the mafic intrusives, it plots within the calc-alkaline field on an AFM diagram, and is characterised by higher mean SiO_2, Al_2O_3, Na_2O, K_2O, Ba and Rb, and lower mean MgO, CaO, Cu, Cr, Ni and V than the mafic intrusives. Whether it represents a stage of calc-alkaline intrusive activity within the Complex postdating the emplacement of the Dark Gneiss is hard to conclude because it is the only occurrence of this type exposed and its contact relationships are unknown. Although dioritic bodies of late Precambrian age occur in both Wales and England, the early age of intrusion of the Greenore Diorite predating the onset of the earliest metamorphism in the Rosslare Complex, suggests that it is unlikely to form part of the same suites.

4.2.2 The second event (M2)

The second tectono-thermal event has been dated. $^{40}Ar/^{39}Ar$ dating from amphibole separates from Dark Gneiss in the St. Helen's area yielded an age of 626 ± 6 Ma, which has been interpreted as representing the peak of the second metamorphic event (Max and Roddick, 1989). A slightly younger date of 618 ± 6 Ma was obtained from the St. Helen's Gabbro, which was interpreted as a synmetamorphic intrusion during the second event (Max, 1975). During this event mid-amphibolite facies conditions prevailed. The St. Helen's Gabbro lacks the coarse gneissose textures of the Greenore Diorite but amphibolite facies conditions are indicated by the replacement of a gabbroic mineral texture by red garnet, quartz, oligoclase and green amphibole. The margins of the gabbro crosscut an earlier foliation within the adjacent gneisses in which green amphibole (M2) overprints the earlier brown amphibole (M1). In the Grey Gneiss at Kilmore Quay rarely preserved M1 kyanite was almost completely pinitised during M2, while in the same area F2 folding of F1 gneissose banding is common.

Mylonites along the northern margin of the Complex which deform the gneissose banding, have sinistral shear bands (Murphy, 1990) and could be related to the second event affecting the gneisses. The well constrained ages of this event suggest that it is a Cadomian or Monian event.

4.2.2.1 The St. Helen's Gabbro
The St. Helen's Gabbro, intruded at the peak of the late Precambrian Cadomian metamorphism, was emplaced at broadly the same time as similar intrusive bodies in England and Wales, such as the Malvern Diorite and the Johnstone Complex. However, whereas the

latter are both reported to have a calc-alkaline chemistry (Thorpe, 1972; Thorpe *et al.*, 1984), both major and trace element concentrations in the St. Helen's Gabbro confirm its essentially tholeiitic gabbroic composition. On a Zr/TiO_2–Nb/Y discriminant plot it emerges as a mildly subalkaline gabbro, while high average concentrations of TiO_2, P_2O_5, Nb, V, Y and LREE suggest that the magma was already quite highly evolved when emplaced. The rocks with the highest concentrations of these elements also contain high Fe_2O_3 (total), suggesting that they fall on a typical tholeiitic iron-enrichment trend. This tholeiitic affinity is confirmed by a Y–FeO (total)/MgO discriminant diagram (Figure 4.2D).

4.2.2.2 Older basic dykes   Following the second event basic dykes were intruded throughout the Complex. These 'Older Basic Dykes' usually lie parallel to the gneissose foliation, but sometimes crosscut at angles of up to 30°. In the Kilmore Quay Section, bifurcation of dykes occurs and some of the dykes crosscut F2 fold axes in the gneisses. No relic igneous minerals have been seen in these dykes; they are thoroughly metamorphosed.

The composition of the Older Basic Dykes is more uniform than the Dark Gneiss, and they tend to form a compact group on most discriminant plots. They show the low Zr/TiO_2 and Nb/Y of subalkaline basalts (Figure 4.2B) and plot in the tholeiitic field on a Y–FeO (total)/MgO variation diagram (Figure 4.2D). They possess a low $(Ce/Y)_{cn}$ ratio suggesting a flat REE profile, and are characterized and distinguished from other basic rocks in the Rosslare Complex by relatively low concentrations of TiO_2, MnO, P_2O_5, Nb, Y, Zn and LREE and moderately enhanced Cr and Ni. All the dykes possess rather low FM values and thus, according to the tholeiite classification devised by Wood (1978) were probably high-magnesia basalts. However, values of SiO_2, K_2O and Rb are all higher than in normal fresh high-magnesia basalts, suggesting metasomatic addition of these elements during metamorphism.

4.2.3   The third event (M3)

In the third event, randomly oriented actinolitic amphibole dated at 480 Ma (Max and Roddick, 1989) and epidote replaced both the brown and green amphiboles in the Dark Gneiss and the Older Basic Dykes, especially around Kilmore Quay where garnet is replaced by an aggregate of epidote, chlorite and quartz. In the St. Helen's area chlorite and quartz replace garnet. Large grains of muscovite and fresh brown biotite overgrow S3 strain-slip fabrics in the Kilmore Quay area. Pervasive F3 folding accompanying this upper greenschist metamorphism deformed both the gneisses and the Older Basic Dykes producing a composite axial planar S3 schistosity. This S3 schistosity is developed in both the gneisses and the Older Basic Dykes. In the Dark Gneiss S3 is a strain-slip schistosity, whereas in the Older Basic Dykes it becomes a fine, penetrative S1 schistosity, which contains a subhorizontal stretching lineation and sinistral shear bands. In the gneisses, by contrast, there is also a pronounced L3 lineation.

In the Ballycogly shear zone $S2_{myl}$ foliation is developed indicating dip-slip movement, which is coeval with the development of weak cleavage in the Arenig Tagoat Group. This field relationship is consistent with a post-Arenig age for this event of 480 Ma.

4.2.4   The fourth event (M4)

In the fourth event greenschist facies retrogression occurred within shear zones, indicated by the growth of secondary muscovite. Minor shearing also affected the margins of the Saltees Granite (intrusion date 436 ± 7 Ma; Max *et al.*, 1979), but did not affect the Carnsore Granite (intrusion date 432 ± 3 Ma; O'Connor *et al.*, 1988) and hence this event can be dated as early Silurian with some precision.

4.3   Rocks marginal to the Rosslare Complex

Within the shear zones at the margins of the Rosslare Complex occur mylonitized gneisses and deformed sedimentary rocks (Figure 4.1). These include, on the northern margin:

4.3.1   Tomhaggard Zone

This zone consists of polydeformed mylonitized quartzofeldspathic gneiss and mylonitized amphibolite which Murphy (1990) considered to have a gabbroic protolith similar to the St. Helen's Gabbro. The retention of a gneissic protolith in the Tomhaggard Zone suggests derivation from the Rosslare Complex (Max, 1975; Max and Long, 1985).

4.3.2   Silverspring Zone

Situated immediately north of the Tomhaggard Zone, the Silverspring Zone consists predominantly of an undated metasedimentary sequence of fine- to coarse-grained greywackes, siltstones and conglomerates, although some gneisses may be tectonically interleaved. Much of the clastic material in the metasediments was derived from a previously mylonitized gneissose quartzo-feldspathic source (Murphy, 1990). A slaty cleavage is similar in orientation to the $S2_{myl}$ foliation in the adjacent mylonites. Murphy (1990) concluded that the Silverspring sediments developed as a cover sequence and were first deformed during $D2_{myl}$ deformation of the adjacent mylonites.

Table 4.2 Mean values and standard deviations for the major component rock suites in the Rosslare Complex. Data from the Tuskar Metavolcanics (from Max and Ryan, 1986) are included for comparative purposes

	Rosslare Complex													
	Grey Gneisses													
	St. Helen's		Kilmore Quay		Dark Gneiss		Greemore Diorite		St. Helen's Gabbro		Older Basic Dykes		Younger Basic Dykes		Tuskar Metavolcanics	
	$\bar{x}$	s	$\bar{x}$	s	$\bar{x}$	s	$\bar{x}$	s	$\bar{x}$	s	$\bar{x}$	s	$\bar{x}$	s	$\bar{x}$	s
SiO_2	64.73	6.41	64.23	1.84	45.62	3.82	54.25	0.39	47.30	1.89	50.14	0.66	48.24	0.73	47.15	0.63
TiO_2	0.97	0.31	0.97	0.17	2.47	0.97	1.29	0.04	2.65	0.54	1.04	0.24	2.05	0.59	3.67	0.76
Al_2O_3	16.21	3.22	15.49	1.70	14.19	2.70	18.29	0.23	15.97	2.03	16.86	1.61	15.10	1.15	14.91	1.09
Fe_2O_3	7.28	1.91	7.05	0.53	2.95	2.48	3.89	0.18	5.41	2.83	3.03	0.44	4.52	1.32		
FeO					8.69	2.33	5.69	0.52	8.11	1.81	5.04	1.04	7.29	1.45	13.16	0.36
MnO	0.10	0.03	0.09	0.02	0.21	0.08	0.21	0.04	0.24	0.06	0.15	0.01	0.22	0.03	0.35	0.04
MgO	2.09	0.43	2.16	0.36	7.61	2.72	3.47	0.07	6.07	0.82	6.79	1.22	6.77	1.24	7.21	0.92
CaO	1.41	0.73	2.40	1.66	7.04	2.78	5.35	0.21	7.45	2.17	9.59	0.69	9.58	1.01	10.36	1.30
Na_2O	2.30	0.97	2.55	0.83	1.73	0.87	3.26	0.38	2.03	0.27	2.54	0.39	2.26	0.46	2.67	0.27
K_2O	2.75	1.27	3.02	0.65	0.83	0.56	1.40	0.46	0.87	0.23	1.02	0.28	0.50	0.20	0.59	0.22
P_2O_5	0.25	0.32	0.13	0.02	0.42	0.31	0.28	0.04	0.53	0.46	0.22	0.12	0.32	0.20	0.26	0.07
H_2O^+	2.19	0.90	1.77	0.17	6.73	5.10	2.85	0.13	2.87	0.88	2.68	0.58	2.55	1.26	1.40	0.45
S	0.01	0.01	0.01	0	0.02	0.01	0.01	0.01	0.03	0.02	0.01	0.01	0.02	0.01	—	

Cl	465	296	427	166	817	358	809	211	690	203	933	318	625	210	—	—
Ba	588	274	603	138	180	111	368	127	289	147	164	89	99	56	—	—
Ce	106	102	47	4	37	36	27	10	44	33	8	8	17	27	21	6
Cr	72	29	69	12	177	206	2	1	100	76	234	125	194	48	—	15
Cu	16	12	37	27	64	28	23	3	105	79	39	27	43	16	188	—
Ga	20	10	19	3	21	5	20	2	24	4	20	3	20	3	—	3
La	41	32	27	10	20	15	13	5	24	14	7	10	7	11	4	3
Nb	14	6	15	2	21	14	10	1	24	13	5	2	10	8	9	6
Nd	52	43	35	9	32	20	11	2	33	19	11	8	15	12	17	22
Ni	33	9	26	2	100	105	9	1	72	29	102	77	72	30	66	—
Pb	17	7	23	8	14	4	13	1	16	3	15	3	17	5	—	7
Rb	103	43	122	16	33	27	52	22	26	8	38	10	23	11	19	28
Sr	187	69	170	94	212	143	334	13	383	108	375	68	283	114	180	—
Th	11	8	11	2	2	2	2	1	3	2	1	2	1	1	—	54
V	125	51	111	8	380	130	157	6	412	112	230	34	328	60	332	6
Y	65	73	34	5	41	21	33	2	47	27	22	3	40	11	23	4
Zn	94	35	81	11	113	28	105	10	131	29	84	18	106	28	93	40
Zr	237	38	198	29	260	232	164	22	183	40	119	36	169	65	113	
n =	6		4		13		3		8		6		8	5	5	

4.3.3   Tagoat Group

Separated from the Silverspring Zone by a 3 km-wide area devoid of outcrop, the Tagoat Group occurs along strike, cropping out in two faulted outliers at Rosslare Harbour and at Tagoat, west of Rosslare Harbour (Figure 4.1). The Group consists of c. 1500 m of weakly deformed, unfossiliferous quartz arenite, conglomerate, red and grey siltstone and shale, overlain with apparent conformity by marine siltstone and turbiditic greywackes which yield shallow water Upper Arenig faunas (Brenchley et al., 1967). Clasts in the basal conglomerate are similar to mylonitic rocks of the Tomhaggard Zone: hence these sediments may have been deposited unconformably on the mylonite zone.

4.3.4   The Ballycogly Group

Consisting of thinly layered siliceous and micaceous phyllonites up to 1300 m in tectonic thickness, this group displays similarities in deformational style and history with the Tomhaggard Zone. However, its derivation from a metasedimentary protolith argues against its inclusion within the Rosslare Complex and Murphy (1990) has suggested that it was a late Precambrian succession. It is not known whether this succession could have been cover to the Rosslare Complex.

4.3.5   Cullenstown Formation

The Cullenstown Formation (Figure 4.1) is an undated sequence of poly-deformed quartzites and metagreywackes which are locally mylonitic and were considered to be late Precambrian by Max and Dhonau (1974). More recently the Formation has been equated with the Bray Group in the Leinster Massif (Bruck et al., 1979) which has been dated on micropalaeontological evidence as middle-Lower to lower-Middle Cambrian. However, we consider it to be part of a late Precambrian to Cambrian basin margin sequence that has a different tectonothermal history from the Bray Group. In our view it therefore differs from the Bray Group, even though it may once have formed part of the same sedimentary prism. It is everywhere separated from the Bray Group by tectonic boundaries and appears to have been subjected to a slightly higher grade of metamorphism.

4.3.6   Tuskar Group

The Tuskar Group (Max and Ryan, 1986) is exposed on the Tuskar Rock and adjacent submarine outcrops (Max and Ryan, 1986). It probably includes the St. Margaret's Gneisses (Baker, 1970; Max, 1975), identified as a dark rock with a simple foliation occurring along the line of the Kilmore-Wilkeen shear zone (Figure 4.1) annealed by the Saltees Granite (Max and Ryan, 1986), and south of the shear zone between St. Helen's and Carnsore Point (Figure 4.1). At the type locality the Tuskar Group is an undated, low grade sequence of metasediments and metavolcanics displaying a single cleavage, which was intruded and hornfelsed by the Carnsore Granite. The basic metavolcanics in the Tuskar Group have been analysed (Max and Ryan, 1986) (Table 4.2) and, although tholeiitic (Figure 4.2D), bear scant chemical resemblance to any of the basic rocks within the Rosslare Complex. In particular they differ from the metabasic rocks of the Rosslare Complex in possessing higher TiO_2, MnO, Ce, Cu and Nd, and lower P_2O_5, Zr/TiO_2, Zr, Sr and Y (Figure 4.2B and C).

4.4 Caledonian intrusions

4.4.1 Saltees Granite

The albite-quartz-microcline Saltees Granite (436 ± 7 Ma, Rb–Sr whole rock; Max et al., 1979) has a foliated northwest margin which anneals a major D3 NE–SW trending sinistral shear zone (Murphy, 1990). This foliation traces into an apparent S3 fabric in the adjacent gneisses, but it is the second deformation in the Older Basic Dykes which were intruded after D2 in the Rosslare Complex Gneisses. We interpret the early deformation of the Older Basic Dykes as D3; hence their later deformation and that of the Saltees Granite should be regarded as D4. Within the granite early albite is strongly deformed and partially replaced by unstrained younger microcline. This D4 strain is not shared by the Younger Basic Dykes, which intrude both the granite and the rest of the Complex.

4.4.2 Younger basic dykes

These dykes are interpreted as being intruded in the final stages of crystalliz-ation of the Saltees Granite because a set of granophyric pipe structures exposed on the north shore of the Greater Saltee (Figure 4.1) indicate fluid interaction between younger basic dyke magmas and the granite. Both within the Saltees Granite and elsewhere in the Rosslare Complex and the Ballycogly Group (Murphy, 1990), green actinolitic amphibole, oligoclase and quartz form mottled intergrowths within the Younger Basic Dykes and suggest that greenschist facies conditions prevailed, possibly during very slow cooling of the Complex. Chemical studies of these dykes show that they are tholeiitic in composition. However, possibly because of the small sample set, they divide into two chemical groupings: those with higher Zr, Y, Nb, P_2O_5 and TiO_2, and those with lower concentrations of these elements. It is not at present clear, therefore, whether more than one suite of Younger Basic Dykes exists. Although these dykes could conceivably be comparable in age to the

Tuskar Group volcanics the numerous chemical differences suggest that they are unlikely to belong to the same suite.

4.4.3 Carnsore Granite

Finally, the post-tectonic Carnsore Granite was intruded at the southeastern margin of the Complex. Originally dated from a Rb–Sr whole-rock mineral isochron at about 550 Ma (Leutwein et al., 1972), recent work has yielded ages of 432 ± 3 (U–Pb) and 428 ± 11 Ma (whole-rock Rb–Sr; O'Connor et al., 1988). The Carnsore Granite is apparently not intruded by Younger Basic Dykes and provides a minimum age for both the deformation of the Tuskar Group and the end of the fourth metamorphic event in the Rosslare Complex.

4.5 Conclusions

The general conclusions which can be reached about the Rosslare Complex are:

(1) It is a tectonically isolated, wedge-shaped block of Proterozoic continental basement in southeast Ireland, which occurs along the regional strike to the southwest of the Mona Complex in Anglesey.

(2) It is flanked by broad mylonitized zones which record sinistral and vertical shear of probable late Precambrian Cadomian and early Palaeozoic age. Sinistral motion of broadly comparable age has also been recorded from the Menai Straits Zone southeast of the Mona Complex in Anglesey (Gibbons, 1987).

(3) Reworking of the gneisses within the Complex during the formation of mylonitic zones confirms the existence of older, pre-Cadomian basement, indicated from isotopic analyses.

(4) The various mafic suites within the Complex are unrelated to each other, and mark separate episodes in the long history of the Complex. None of these suites is related to the volcanics within the Tuskar Group. Hence there are no obvious geological links between the Tuskar Group and the older Rosslare Complex, which might therefore belong to separate terranes.

(5) The Rosslare Complex may represent a suspect terrane accreted onto the margin of the Avalonian Superterrane during the late Precambrian.

(6) The 'docking' of the Rosslare terrane with the Avalonian Superterrane is associated with the formation of the mylonitic zones, but final locking of these zones did not take place until the intrusion of the Saltees Granite. However, Lower Palaeozoic sediments within these zones suggest that docking was pre-Arenig, and possibly as early as the late Precambrian.

(7) As a suspect terrane the Rosslare Complex remains an enigma: its former relationships (if any) with gneisses of similar age in Brittany and the Channel Islands remain unresolved.

References

Baker, J. W. (1966) The Ordovician and other post-Rosslare series rocks in SE Co. Wexford. *Geol. J.* 5, 1–6.

Baker, J. W. (1970) Petrology of the metamorphosed Precambrian rocks of southeasternmost Co. Wexford. *Proc. R. Irish Academy*, 69, 1–20.

Barber, A. J., Max, M. D. and Bruck, P. M. (1981) Field meeting in Anglesey and southeastern Ireland, 4–11 June 1977. *Proc. Geol. Assoc.* 92, 269–291.

Brenchley, P. J., Harper, J. C. and Skevington, D. (1967) Lower Ordovician shelly and graptolitic faunas from SE Ireland. *Proc. R. Irish Academy* 65, 385–390.

Bruck, P. M., Colthurst, J. R. J., Feely, M., Gardiner, P. R. R., Penney, S. R., Reeves, T. J., Shannon, P. M., Smith, D. G. and Vanguestaine, M. (1979) Southeast Ireland: Lower Palaeozoic stratigraphy and depositional history. In *The Caledonides of the British Isles—Reviewed*, eds. Harris, A. L., Holland, C. H. and Leake, B. E., *Spec. Publ. Geol. Soc. London* 8, 533–544.

Calloway, C. (1881) The metamorphic and associated rocks south of Wexford. *Geol. Mag.* 8, 1–5.

Davies, G., Gledhill, A. and Hawkesworth, C. (1985) Upper crustal recycling in southern Britain: evidence from Nd and Sr isotopes. *Earth Planet. Sci. Lett.* 75, 1–12.

Floyd, P. A. and Winchester, J. A. (1978) Identification and discrimination of altered and metamorphosed volcanic rocks using immobile elements. *Chem. Geol.* 21, 291–306.

Gibbons, W. (1987) Menai Strait Fault System: an early Caledonian terrane boundary in North Wales. *Geology* 15, 744–747.

Leutwein, F., Sonet, J. and Max, M. D. (1972) The age of the Carnsore Granodiorite. *Bull. Geol. Surv. Ireland* 1, 303–309.

Max, M. D. (1975) Precambrian rocks of SE Ireland. In *A Correlation of Precambrian Rocks of the British Isles*, eds. Harris, A. L. et al., *Spec. Rep. Geol. Soc. London* 6, 97–101.

Max, M. D. (1978) Day 1. The Rosslare District: Pre-Caledonian rocks, late Precambrian and Cambro-Ordovician sediments, Carnsore and Saltees Granites. In *Field Guide to the Caledonian and Pre-Caledonian Rocks of Southeast Ireland*, eds. Bruck P. *et al.*, *Geol. Surv. Ireland Guide Ser.* 2, 15–23.

Max, M. D. and Dhonau, N. B. (1971) A new look at the Rosslare Complex. *Proc. R. Dublin Soc.* 4A, 103–120.

Max, M. D. and Dhonau, N. B. (1974) The Cullenstown Formation: Late Precambrian sediments in southeast Ireland. *Geol. Surv. Ireland Bull.* 1, 447–458.

Max, M. D. and Long, C. B. (1985) Pre-Caledonian basement in Ireland and its cover relationships. *Geol. J.* 20, 341–366.

Max, M. D. and Roddick, J. C. (1989) Age of metamorphism in the Rosslare Complex, southeast Ireland. *Proc. Geol. Assoc.* 100, 113–121.

Max, M. D. and Ryan, P. D. (1986) The Tuskar Group of southeastern Ireland: its geochemistry and depositional provenance. *Proc. Geol. Assoc.* 97, 73–79.

Max, M. D., Ploquin, A. and Sonet, J. (1979) The age of the Saltees Granite in the Rosslare Complex. In *The Caledonides of the British Isles—Reviewed*, eds. Harris, A. L., Holland, C. H. and Leake, B. E., *Spec. Publ. Geol. Soc. London* 8, 723–725.

Murphy, F. C. (1990) Basement-cover relationships of a reactivated Cadomian mylonite zone, Rosslare Complex, SE Ireland. In *The Cadomian Orogeny*, eds. D'Lemos, R. S., Strachan, R. A. and Topley, C. G., *Spec. Publ. Geol. Soc. London* 51, 329–339.

O'Connor, P. J., Kennan, P. S. and Aftalion, M. (1988) New Rb–Sr and U–Pb ages for the Carnsore Granite and their bearing on the antiquity of the Rosslare Complex, southeastern Ireland. *Geol. Mag.* 125, 25–29.

Pearce, J. A. (1979) Geochemical evidence for the genesis and eruptive setting of lavas from Tethyan Ophiolites. In *Proceedings of the International Ophiolite Symposium, Cyprus*, ed. Panayiotou, A., 261–272.

Thorpe, R. S. (1972) Possible subduction zone origin for two Precambrian calc-alkaline plutonic complexes from southern Britain. *Bull. Geol. Soc. Am.* 83, 3663–3668.

Thorpe, R. S., Beckinsale, R. D., Patchett, P. J., Piper, J. D. A., Davies, G. R. and Evans, J. A. (1984) Crustal growth and late Precambrian–Early Palaeozoic plate tectonic evolution of England and Wales. *J. Geol. Soc. London* 141, 521–536.

Williams, H. and Max, M. D. (1980) Zonal subdivision and regional correlation in the Appalachian–Caledonide Orogen. In *The Caledonides in the USA*, ed. Wones, D. R., Virginia Polytechnic Inst. Mem. 2, 57–62.

Winchester, J. A. and Floyd, P. A. (1977) Geochemical discrimination of different magma series and their differentiation products using immobile elements. *Chem. Geol.* **20**, 325–343.

Winchester, J. A. and Max, M. D. (1982) The geochemistry and origins of the Precambrian rocks of the Rosslare Complex, southeast Ireland. *J. Geol. Soc. London* **139**, 309–319.

Wood, D. A. (1978) Major and trace element variations in the Tertiary lavas of Eastern Iceland, and their significance with respect to the Iceland geochemical anomaly. *J. Petrol.* **19**, 393–436.

5 Cadomian terranes in the North Armorican Massif, France

R. A. STRACHAN, R. A. ROACH and P. J. TRELOAR

5.1 Introduction

Of the several isolated massifs which expose pre-Mesozoic rocks in Central and Western Europe, the Armorican Massif (Figure 5.1) is of particular interest since a clear insight can be obtained into the evolution of the late Precambrian Cadomian belt in the northern part of the Massif. Here, with the exception of northwest Brittany, there was no major tectonothermal activity during the Palaeozoic Variscan cycle. The Cadomian Orogeny was first defined by Bertrand (1921) as the late Precambrian orogeny which in the North Armorican Massif resulted in the folding, uplift and peneplanation of Brioverian metasediments prior to the deposition of early Palaeozoic red beds.

The Cadomian belt of the North Armorican Massif is mainly composed of Upper Proterozoic supracrustals of the Brioverian succession (Barrois, 1895, 1896, 1908; Graindor, 1957; Cogné, 1962; Le Corre, 1977; Chantraine et al., 1982; Rabu et al., 1982, 1983). Pre-Cadomian Icartian basement gneisses are exposed on Guernsey and Sark, at Cap de la Hague, and as rafts in the Cadomian Perros-Guirec complex of North Brittany (Figure 5.1). A prolonged history of subduction-related calc-alkaline magmatism spans the period 700–500 Ma (Graviou et al., 1988; Brown et al., 1990 and references therein). Main phase deformation and metamorphism occurred in North Brittany between 590 and 540 Ma (Brun and Balé, 1990; Strachan and Roach, 1990; Treloar and Strachan, 1990), prior to uplift, erosion and deposition of early Palaeozoic red bed sequences. There is general agreement that the Cadomian belt is subdivided by steep shear zones and faults into a number of tectonic blocks with contrasting histories (Cogné, 1962; Cogné and Wright, 1980; Dissler et al., 1988; Chantraine et al., 1988; Rabu et al., 1990; Strachan et al., 1989), although the palaeogeographic significance of each of these blocks and the timing and mechanism of their juxtaposition are controversial.

Strachan et al. (1989) have proposed that the Cadomian belt of North Armorica comprises three main tectonic units; the St. Brieuc, St. Malo and Mancellian terranes, which together form the North Armorican composite terrane (NACT) (Figure 5.1). These terranes record the development, defor-

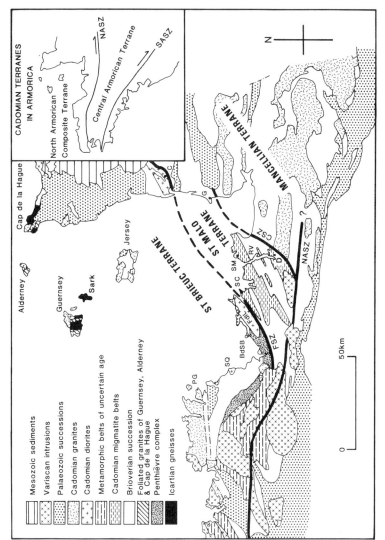

Figure 5.1 Geological map of North Armorica. NASZ = North Armorican shear zone, FSZ = Fresnaye shear zone, PG = Perros-Guirec complex, SQ = St. Quay basic intrusion, BdSB = Baie de St. Brieuc, FdlL = Fort de la Latte quartz diorite, SM = St. Malo, SC = St. Cast, D = Dinan, RV = Rance Valley, G = Granville, C = Coutances, SASZ = South Armorican shear zone. Modified from Strachan *et al.* (1989).

mation and accretion of continental Cadomian arcs and marginal basins along an active plate margin probably located to the north or northwest of the Armorican Massif. The NACT is separated from the Central Armorican Terrane to the south (Figure 5.1) by the North Armorican shear zone, a Cadomian lineament reactivated during the Variscan (Watts and Williams, 1979). In the Central Armorican Terrane, although sediments which have been assigned to the Brioverian are separated from their Palaeozoic cover by an angular unconformity they were not subject to major orogenic activity until the Variscan (Gapais and Le Corre, 1980; Hanmer *et al.*, 1982). In this contribution we describe the geology of the best exposed segment of the Cadomian belt, namely North Brittany, within the framework of three geologically distinct terranes which were juxtaposed during sinistral transpression at *c.* 540 Ma.

5.2 St. Brieuc terrane (SBT)

The SBT (Figure 5.1) is the most lithologically variable of the three terranes and includes a wide range of Cadomian gabbroic, dioritic and granitic calc-alkaline complexes emplaced into both Proterozoic basement gneisses and Brioverian volcano-sedimentary successions during the period 700–500 Ma (Vidal *et al.*, 1972, 1974; Auvray, 1979; Graviou and Auvray, 1985; Guerrot and Peucat, 1990).

In North Brittany Cadomian deformation has resulted in the internal imbrication of the St. Brieuc Terrane into five tectonic blocks of differing deformation intensities and metamorphic grade (Figure 5.2). These are mainly bounded by ductile shear zones formed during Cadomian transpression and subsequently reactivated as late Cadomian (and Variscan?) brittle faults. From north to south these blocks are:

(1) *The Trégor block*, formed of a low grade and weakly to moderately deformed Brioverian metavolcanic and metasedimentary sequence, intruded in the north by the Perros Guirec granitoid complex which contains rafts of pre-Cadomian Icartian basement. The Brioverian is overlain unconformably by early Palaeozoic red beds.

(2) *The Kérégal block*, containing a high grade and structurally complex supracrustal assemblage of possible Brioverian affinities, intruded by a suite of variably deformed and metamorphosed granitoids.

(3) *The Binic block*, formed of weakly deformed, greenschist facies Brioverian turbidites.

(4) *The Roselier block*, incorporating strongly deformed, upper greenschist to amphibolite facies Brioverian metavolcanics and metasediments.

(5) *The Plurien block*, formed of orthogneisses and calc-alkaline granitoids of the early Cadomian Penthièvre complex, overlain unconformably by low

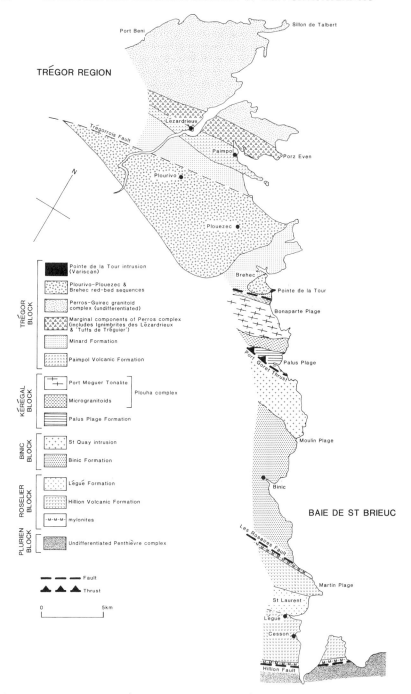

Figure 5.2 Geological map of the area between the Trégor region and the Baie de St. Brieuc, North Brittany.

grade and weakly deformed Brioverian volcanics and intruded by the Fort de la Latte quartz diorite. Both the Brioverian and the Fort de la Latte quartz diorite are overlain by early Palaeozoic red beds.

The SBT incorporates three major tectonostratigraphic units, namely pre-Cadomian basement, the early Cadomian Penthièvre complex and Brioverian supracrustals, each of which will be described within the tectonic framework outlined above.

5.2.1   Pre-Cadomian Icartian basement

Relics of a pre-Cadomian basement, of probable Gondwanan affinities, crop out as enclaves or rafts within the Cadomian Perros-Guirec plutonic complex of the Trégor block (Figure 5.2). At Port Beni (Figure 5.2) foliated, migmatitic augen gneisses of granodioritic composition, with associated metatexites and amphibolite sheets, occur as detached rafts within an unfoliated Cadomian host, the Pleubian microgranodiorite (Figure 5.2) (Auvray, 1979). A U–Pb zircon age of 1790 ± 10 Ma has been obtained from the Port Beni augen gneisses (Auvray et al., 1980a) prompting correlation with the Icart gneisses of Guernsey (see Section 5.7.1). Auvray et al. (1980a) have suggested that large areas of basement gneisses are also present in the west of the Trégor region, represented by the Moulin de la Rive othogneiss complex of Griffiths (1985) (Figure 5.3) which has yielded U–Pb zircon ages of 2000 and 2013 Ma. Griffiths (1985) has, however, argued that the majority of the Moulin de la Rive orthogneiss complex comprises highly strained members of the Cadomian Perros-Guirec granitoid complex. According to Griffiths (1985) the zircons dated are either xenocrysts or more probably are derived from small areas of augen gneiss which may represent highly deformed rafts of basement similar to those dated at Port Beni.

5.2.2   The Penthièvre complex

The Penthièvre complex, which crops out along the east side of the Baie de St. Brieuc within the Plurien block (Figures 5.2 and 5.4), consists of variably deformed and metamorphosed amphibolitic, granodioritic, trondhjemitic and quartz-dioritic orthogneisses, intruded by a suite of quartz diorites and granodiorites (Shufflebotham, 1990). The complex is unconformably overlain along its western margin by the local Brioverian succession (Cogné, 1959) and was formerly thought to represent the type area of a pre-Cadomian 'Pentevrian' basement. Early isotopic work yielded K–Ar and Rb–Sr mineral and whole-rock ages of c. 1200–900 Ma (Leutwein et al., 1968), prompting correlation with Grenvillian-Sveconorwegian events (Roach et al., 1972). Recent isotopic work suggests that the Penthièvre complex is both substantially younger than previously thought and composite in nature. Guerrot

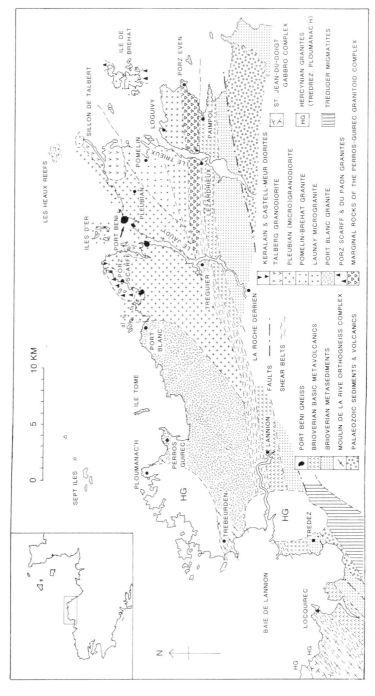

Figure 5.3   Geological map of the Trégor region (modified from Auvray, 1979 and Griffiths, 1985).

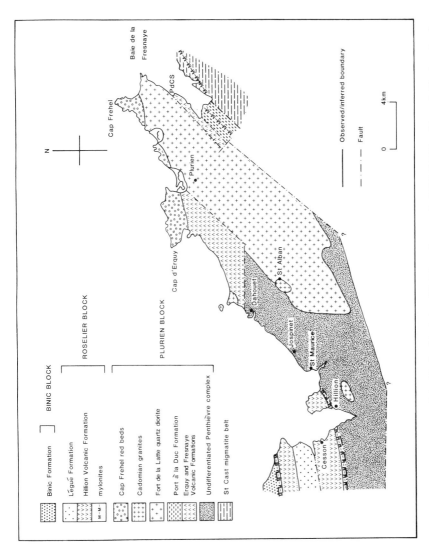

Figure 5.4   Geological map of the Plurien block (modified from Lees *et al.*, 1987). PdCS = Plage de Chateau Serein.

and Peucat (1990), report U–Pb zircon ages of 667 ± 4 and 656 ± 5 Ma from foliated granitoid boulders within the Brioverian Cesson conglomerate (see below) of the adjacent Roselier block (Figure 5.2). This conglomerate is generally thought to have been derived from the erosion of the orthogneisses and granitoids which form the western part of the Penthièvre complex (Cogné, 1959; Rabu et al., 1982, 1983; Roach et al., 1986). We interpret this part of the Penthièvre complex as representing a fragment of an early Cadomian volcanic arc, a proposal consistent with its calc-alkaline geochemistry (Shufflebotham, 1989). The poorly exposed Yffiniac metagabbros which lie along the southern margin of the complex have yielded U–Pb zircon ages of c. 602 Ma (Balé and Brun, 1983) and may have been intruded into the remnant arc during main-phase Cadomian magmatism.

5.2.3 Brioverian supracrustals

The base of the Brioverian succession in the Baie de St. Brieuc is probably represented in the Roselier block by a thick sequence of strongly deformed, low-amphibolite facies metabasalts with metasedimentary interbeds, the Hillion Volcanic Formation (HVF) (Lees et al., 1987; Roach et al., 1990). At Cesson the base of the HVF is marked by a conglomeratic clastic sequence which includes pebbles of granodiorite, quartz-diorite, amphibolite, granitic gneiss and leucogranite probably derived from the erosion of the Penthièvre complex to the southeast. The basal metasediments of the HVF pass rapidly northwards into a c. 2 km-thick sequence of massive, locally pillowed, tholeiitic metabasalts, with thin interlayers of psammite and pelite (Roach et al., 1990). The sequence is repeated to the north at Martin Plage (Figure 5.2) on the north limb of a major Cadomian fold, the St. Laurent synform (Rabu et al., 1982, 1983; Strachan and Roach, 1990). The HVF is overlain by the Legué Formation which comprises a sequence of thinly interbanded pelites, semi-pelites and psammites, with occasional calc-silicate concretions.

The Brioverian rocks of the Plurien block to the southeast (Figure 5.4) comprise basic volcanics of the Erquy and Fresnaye formations which are overlain in the Baie de la Fresnaye by the pelites and psammites of the Port à la Duc Formation (Figure 5.4). Both volcanic formations have been correlated with the HVF of the Roselier block (Lees et al., 1987; Strachan and Roach, 1990) and the Port à la Duc Formation may be equivalent to the lithologically similar Legué Formation. Metasedimentary units of the Erquy Volcanic Formation unconformably overlie the Penthièvre complex at Jospinet (Figure 5.4) (Cogné, 1959; Roach et al., 1986, 1988). Largely undeformed metadolerites (Dahouet dykes) which intrude the Penthièvre complex on the west side are inferred by Lees et al. (1987) to represent feeders to the Brioverian basic volcanics. The HVF lies in tectonic contact north of Martin Plage (Figure 5.2) with the Binic block. This includes low- to mid-

Figure 5.5   Brioverian turbidites of the Binic Formation, Binic, St. Brieuc terrane.

greenschist facies sandstones and siltstones of the Binic Formation (Figure 5.5) (Barrois, 1908; Jeanette, 1971; Ryan, 1973; Rabu *et al.*, 1982, 1983). Although the contact between the HVF and the Binic Formation is tectonic, most workers consider the Binic Formation to be the younger of the two units (Strachan and Roach, 1990), and that it may in part correlate with the Legué Formation (e.g. Rabu *et al.*, 1982, 1983).

The volcano-sedimentary association of basic volcanics and turbidites also occurs in the Trégor block in the north of the SBT (Figure 5.3). The structurally lowest Brioverian rocks in this area are a thick succession of basic metavolcanics which strike E–W through the Trégor, parallel to the southern margin of the Cadomian Perros-Guirec complex (Figure 5.3). In the east of the Trégor block these are represented by low-greenschist facies metabasalts, occurring as massive and pillowed flows, intrusive sheets and crudely bedded broken pillow breccias (Roach *et al.*, 1988), Known as the Spilites de Paimpol (Barrois, 1908; Auvray, 1979), these rocks are here referred to as the Paimpol Volcanic Formation (Figure 5.2) (Lees *et al.*, 1987). They are probably continuous with the lithologically similar metabasic volcanic Pointe de l'Armorique Formation (Griffiths, 1985) which crops out in the west of the Trégor region (Figure 5.3). Metamorphic grade is similar to that of the Paimpol Volcanic Formation, although deformation is generally more intense. The Paimpol Volcanic Formation is overlain to the south by the Minard Formation which is a series of turbiditic greywacke sandstones and siltstones, identical in most respects to the Binic Formation (Strachan and Roach, 1990). The Minard Formation is probably laterally continuous with the turbiditic Plestin Formation which lies to the southeast of the Pointe de l'Armorique Formation in the west of the Trégor region (Figure 5.3) (Griffiths, 1985).

The gross similarity of Brioverian sequences in the southern part of the Baie de St. Brieuc and the Trégor region prompts correlation of firstly the basic volcanics of the Paimpol and Pointe de l'Armorique formations with those of the Hillion-Erquy-Fresnaye formations, and secondly the turbidites of the Minard and Plestin formations with those of the Binic Formation (Strachan and Roach, 1990). This suggests a sequence of events within the Brioverian basin of the SBT (= Lanvollon basin of Dupret et al., 1990) which involves post-650 Ma rifting of Lower Proterozoic basement and early Cadomian arc complexes and subsequent extrusion of thick sequences of tholeiitic submarine basalts. The abrupt cessation of volcanism was followed by the deposition of turbiditic sediments, probably in a variety of submarine fan environments (Denis and Dabard, 1988).

5.2.4   Cadomian magmatism

A major early phase of intrusive calc-alkaline activity, possibly associated with the inversion of the Brioverian (Lanvollon) basin, is reflected in the emplacement of the Perros-Guirec complex (614 ± 7 Ma; Graviou and Auvray, 1985), the Fort de la Latte quartz-diorite (593 ± 17 Ma; Vidal et al., 1974) and the St. Quay basic intrusion (584 ± 58 Ma; Vidal et al., 1972) (Figure 5.1).

The Perros-Guirec complex is mainly composed of granites and granodiorites with minor diorites and monzonites (Auvray, 1979) (Figure 5.3). A U–Pb zircon age of 614 ± 7 Ma has been obtained from the Talberg granodiorite (Graviou and Auvray, 1985). That the Longuivy microgranite has yielded an Rb–Sr isochron of 544 ± 19 Ma (Vidal, 1980), suggests however, the presence of at least two intrusive suites within the complex. The southern margin of the Perros-Guirec complex is characterised by a belt of porphyritic granitoids which west of Tréguier are deformed in a major high strain zone termed the Locquirec-Tréguier shear belt by Griffiths (1985). The mylonitised porphyritic microgranitoids superficially resemble a sequence of tuffaceous volcanic rocks, the so-called 'Tuffs de Tréguier' or 'Tuffs de Locquirec' (Barrois, 1908; Verdier, 1968; Auvray, 1979; Autran et al., 1979), which have been incorporated as an integral lithostratigraphic unit at the base of the Brioverian in the Trégor region (e.g. Auvray et al., 1976; Auvray, 1979; Rabu et al., 1983). Detailed remapping of the low strain parts of this belt east of Tréguier has shown that these rocks are not primarily tuffaceous and extrusive in origin, but are mainly a sequence of porphyritic intermediate to acid intrusive sheets emplaced along the south margin of the Perros-Guirec complex (Roach et al., 1986, 1988; Strachan and Roach, 1990). The strati-graphic position of the rhyolitic unit known as the 'Ignimbrites de Lezardrieux' thought by Auvray et al. (1976) and Auvray (1979) to be a subhorizontal sequence of Cambrian volcanic rocks which unconformably overlie the steeply dipping 'Tuffs de Tréguier', is less certain. While undoubtedly a series of pyroclastic deposits and lava flows, there is no evidence for an unconformity with the 'Tuffs de Treguier'. Rather, they occur as concordant bands within the largely intrusive sheets forming the 'Tuffs de Tréguier' (Roach et al., 1986, 1988; Strachan and Roach, 1989). Their isochron age of 547 ± 12 Ma (Vidal, 1980) is thought more likely to reflect devitrification of the vitric components during Cadomian metamorphism rather than their eruption. The complex has been affected by low-grade Cadomian metamorphism and the marginal components are cleaved, demonstrating that intrusion pre-dated Cadomian metamorphism and deformation.

Within the Kérégal block a highly deformed and metamorphosed igneous complex, the Plouha complex, intrudes a sequence of high-grade migmatitic gneisses and amphibolites, the Palus Plage Formation (Figure 5.2) (Strachan and Roach, 1990). The Plouha complex, as yet undated, includes a central tonalite, the Port Moguer Tonalite (Ryan, 1973), which was intruded into an envelope of microgranitoids (Figure 5.2). The Palus Plage Formation and the Plouha complex were considered by Roach et al. (1972) and Ryan and Roach (1975) to represent an inlier of pre-Cadomian 'Pentevrian' basement, probably of similar age to the Penthièvre complex. Strachan and Roach (1990) have, however, suggested that the Palus Plage Formation may be high-grade Brioverian, which carries the implication that the Plouha complex is a mid-crustal section through a pre- or syn-tectonic Cadomian igneous complex.

In the absence of reliable whole-rock isochrons and mineral ages which might accurately date Cadomian deformation and metamorphism, considerable importance has been attached by some workers to the parallelism of igneous foliations within the Fort de la Latte quartz-diorite with the regional cleavage within the Brioverian. Balé and Brun (1983) considered that this implied that intrusion of the Fort de la Latte quartz-diorite at 593 ± 17 Ma (U–Pb zircon age; Vidal et al., 1974) occurred during main-phase Cadomian deformation and metamorphism. Strachan and Roach (1990) have, however, argued that parallelism of fabrics cannot by itself be used to demonstrate synchroneity of intrusion and deformation. The timing of deformation within the SBT is therefore poorly constrained, since field relations between deformed Brioverian rocks and dated intrusions are only clear in the vicinity of the late- to post-tectonic St. Quay intrusion which is imprecisely dated at 584 ± 58 Ma (Rb–Sr isochron; Vidal et al., 1972).

Evidence for late phase Cadomian magmatic activity in the SBT is provided by U–Pb zircon ages of 543 ± 7 and 526 ± 6 Ma obtained from granites near Belle Isle-en-Terre (Figure 5.1) (Andriamarofohatra and La Boisse, 1988). The deformed Beg ar Fourm granodiorite in the west Trégor has yielded a U–Pb zircon age of 528 ± 4 Ma (Guerrot and Peucat, 1990), but here the deformation affecting this body is Variscan.

5.3   St. Malo terrane (SMT)

This terrane includes from northwest to southeast the St. Cast, St. Malo and Dinan migmatite belts and narrow intervening strips of greenschist facies Brioverian metasediments (Figure 5.1). The formation and coeval deformation of the migmatite belts is the main event recorded in the SMT which does not contain any indication of the protracted history of calc-alkaline magmatism present within the SBT.

The migmatites comprise complexly deformed, upper amphibolite facies metatexites and diatexites which grade into homogeneous anatectic granitoids, interpreted as convectively homogenised magma diapirs (Brown, 1979). The St. Malo belt displays a marked zonation from metatexites on the margins to a diatexite core, which on the southeast side of the belt has punched through the metatexite envelope to intrude the adjacent metasediments (Brown, 1978). Internal structures indicate that the St. Malo migmatites lie within an asymmetric dome, the result of the south-verging diapiric overturn and uplift during migmatisation (Brun and Martin, 1979). Phase analysis suggests that migmatisation took place at mid-crustal pressures of 4–7 kbar (Brown, 1979).

The Brioverian rocks of the SMT are mainly interlayered pelites, semipelites and psammites which only rarely contain sedimentary structures. Although various workers (e.g. Chantraine *et al.*, 1988) have suggested that these rocks are characterised by the presence of carbonaceous quartzites (phtanites), we do not believe these to be sufficiently common as to be diagnostic of the SMT. The presence of numerous tectonic breaks, the absence of distinctive marker horizons and the lack of younging indicators mean that it is not possible to establish a reliable Brioverian stratigraphy for the SMT. The contacts between the migmatites and the Brioverian metasediments are generally marked by high strain sinistral shear zones which largely obscure the nature of the original relationship between the two groups of rocks. In the Rance Valley, however, Brioverian phyllites pass transitionally northwards into the migmatites of the St. Malo belt (Figure 5.1), supporting the contention of Brun (1977) that the migmatites formed as a result of the partial melting of Brioverian sediments.

Rb–Sr whole-rock isochrons and U–Pb zircon ages indicate that anatexis occurred during the Cadomian orogeny at *c.* 540 Ma (Peucat, 1986). Strachan *et al.* (1989) have suggested that the Brioverian rocks of the SMT were deposited in an extending marginal basin with high heat flow and that inversion of the basin during early stages of deformation allowed temperatures to rise at mid-crustal levels to *c.* 750°C, resulting in partial melting of Brioverian sediments at pressures of *c.* 5 kbar. Geochemically, the migmatites of the St. Malo belt plot at the acid end of a calc-alkaline trend on both AFM and $(Na_2O + K_2O)$ vs SiO_2 diagrams (Brown *et al.*, 1990). Despite the fact that anatectic granitoids of the St. Malo belt are clearly derived from the partial melting of Brioverian sediments, they have no geochemical affinities with 'S-type' granites as defined by White and Chappell (1983). Their calc-alkaline geochemistry reflects in part the arc-derivation of the Brioverian sediments, modified by fractional crystallisation, and is consistent with an overall arc-subduction setting during the Cadomian Orogeny.

5.4   Mancellian terrane (MT)

The poorly exposed rocks of the MT mainly comprise low-grade Brioverian metasediments intruded by the calc-alkaline granites of the Mancellian batholith (Jonin, 1973, 1981) (Figure 5.1). Apart from narrow shear zones which deform several of the Mancellian granites (Jonin, 1981), regional deformation structures are absent and the weak upright folds present in the Brioverian are generally considered to relate to granite emplacement. The MT is thus distinct in several respects from both the SBT and the SMT.

The Brioverian rocks of the MT are mainly low-grade sandstones, siltstones and mudstones with thin conglomerate horizons. Metamorphic grade locally rises to cordierite-bearing assemblages adjacent to the Mancellian granites (Jonin and Vidal, 1975). Sedimentological studies carried out further east in Lower Normandy suggest that these rocks accumulated as a series of turbidites and submarine fan deposits derived from a source area situated to the north or northeast (Chantraine *et al.*, 1982). The presence within conglomerates of clasts of both volcano-plutonic and sedimentary (phtanites) origin has led to the proposal that this source area is represented by the Brioverian supracrustal rocks and Cadomian plutons presently exposed in the SBT and the SMT (Dissler *et al.*, 1988; Chantraine *et al.*, 1988 and references therein), the inferred intra-Brioverian unconformity being obscured by shear zones and faults.

The rocks which form the bulk of the Mancellian batholith are granodiorites (Jonin, 1981). The most common type is a grey granodiorite with minor cordierite which is associated with rare porphyritic granodiorite, leucogranodiorite and quartz-diorite. Many of the complexes are cut by small bosses of leucogranite. Minor basic complexes are associated with some of the Mancellian granites in Lower Normandy (Le Gall and Mary, 1982, 1983). The presence within the grey granodiorites of *c.* 3% each of cordierite and muscovite and abundant micaceous enclaves was thought by Jonin (1981) to indicate an origin by partial melting of Brioverian sediments. Brown *et al.* (1990) have argued that these features are more likely to represent the effects of variable high level contamination by Brioverian metasedimentary material during batholith emplacement. They propose a petrogenetic model in which mantle-derived magma ponds at the base of the crust to allow melting and assimilation, the resultant mixed magma then undergoing fractionation processes during ascent and emplacement at higher crustal levels. U–Pb monazite ages and imprecise Rb–Sr whole-rock isochrons indicate that intrusion of the Mancellian batholith occurred at *c.* 550–540 Ma (Graviou *et al.*, 1988 and references therein).

5.5 Nature and timing of deformation

Regional deformation structures are restricted to the SBT and the SMT, where the most widespread Cadomian structure is an upright E–W or NE–SW trending schistosity or cleavage which is axial planar to open to isoclinal folds. A gently east- or northeast-plunging stretching lineation is frequently defined by mineral aggregates, rodded quartz veins and in meta-volcanics, stretched pillows and amygdales. Cadomian strain is heterogeneous and concentrated in subvertical, upper greenschist–low amphibolite facies shear zones, marked by highly strained, often mylonitic, L-S or L-tectonites (Brown, 1978; Balé, 1986; Strachan and Roach 1990; Treloar and Strachan 1990). The Fresnaye and Cancale sinistral shear zones define the boundaries between the SBT and the SMT and the SMT and MT respectively (Figure 5.6). Other shear zones internally imbricate both the SBT and SMT (Figure 5.6) and separate tectonic blocks of differing metamorphic grade and/or deformation intensities.

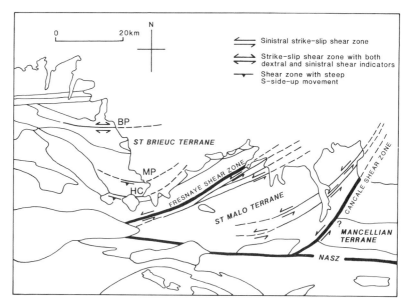

Figure 5.6 Location of Cadomian shear zones and terrane boundaries within North Brittany (BP = Bonaparte Plage shear zone, MP = Martin Plage shear zone, HC = Hillion-Cesson shear zone, NASZ = North Armorican shear zone).

Figure 5.7 Sinistral C-S fabrics within highly deformed Brioverian semipelites. Plage de Pen Guen, 3 km south-southeast of St. Cast, St. Malo terrane.

Lineations and shear criteria, including S–C fabrics, extensional shear bands, rotated and sheared porphyroclasts, displaced passive makers and aligned grains and fibres within quartz veins, give a sub-horizontal generally sinistral strike-slip sense of displacement across the shear zones within and marginal to the SMT (Figures 5.7 and 5.8). Sheath folds within these shear zones indicate simple shear strains of y > 10 (Cobbold and Quinquis, 1980).

Figure 5.8 Subhorizontal stretching and rodding lineation within mylonitised migmatites. Plage de Pen Guen, 3 km south-southeast of St. Cast, St. Malo terrane.

Although sinistral shear criteria are dominant, all shear zones carry some dextral indicators. In some blastomylonites, deformed porphyroclasts are symmetric, indicating a degree of strain partitioning into alternate zones of flattening and simple shearing.

Shear zones within the SBT have a more complex kinematic history. The Bonaparte Plage shear zone, which separates the Trégor and Kérégal blocks (Figure 5.6), contains shear criteria which indicate both dextral and sinistral strike-slip displacements. In contrast, the Martin Plage and Cesson shear zones which, respectively, separate the Binic and Roselier blocks and the Roselier and Plurien blocks (Figure 5.5), both have steeply oblique, S-side-up senses of movement and may represent a mid-crustal analogue of the strike-slip-related 'flower' structures described by Harding (1985). In the Trégor region (Figure 5.3) mylonites of the Locquirec-Treguier strike-slip shear belt are truncated by pre-Variscan basic dykes, suggesting a Cadomian age for the initiation of this structure, although the sense of displacement requires further investigation (Griffiths, 1985).

Treloar and Strachan (1990) have interpreted the Cadomian structures of the SBT and SMT as resulting from transpressive deformation in a dominantly sinistral strike-slip belt. The association of subhorizontal, or gently plunging, stretching lineations with upright planar fabrics implies subhorizontal movement during deformation. Shear criteria indicate that this was achieved by noncoaxial deformation with cross-strike shortening synchronous with strike-slip displacements. The parallelism of planar fabrics to shear zone boundaries, the high shear strains indicated by sheath folds and the predominance of sinistral shear criteria imply that the cumulative displacements across this belt are likely to be considerable.

The Fresnaye shear zone and other shear zones within the SMT contain granodiorite and leucogranite sheets interpreted as late fractions of the migmatitic melt (Figure 5.9) (Brown, 1978; Treloar and Strachan, 1990). These appear to have been emplaced syntectonically into the shear zones since they cut some of the shear fabrics but are deformed by others, suggesting that terrane amalgamation and internal imbrication of the SMT occurred during late stages of migmatisation at c. 540 Ma. The precise age of the shear zones within the SBT is unknown. It is possible that they represent the early stages in a continuous phase of regional transpression which was diachronous to the southeast and culminated in terrane amalgamation at c. 540 Ma. Alternatively, deformation in the SBT may be polyphase: shear zones in the terrane could represent effects of the regional c. 540 Ma terrane amalgamation event superimposed upon fabrics related to a separate, significantly older event.

The interpretation that Cadomian deformation occurred in a sinistrally transpressive, strike-slip setting is at variance with that of Balé and Brun (1983, 1989), Balé (1986) and Brun and Balé (1990) who argue that regional Cadomian deformation was related to southwest-directed thrusting. All the formation boundaries within the SBT between Cesson and Binic are inter-

Figure 5.9   Folded syn-tectonic leucogranite within mylonitised gneisses, Pointe de la Garde, St. Cast-le-Guildo, St. Malo terrane.

preted by them as steep lateral ramps to a series of southwest-directed listric thrusts which crop out inland and root downwards into the 'St. Brieuc Thrust' located along the southern margin of the Penthièvre complex and the Fort de la Latte quartz diorite. Difficulties in accepting this model arise from the almost complete lack of exposure of any of the supposed thrusts. The 'St. Brieuc Thrust' where it is exposed at Plage de Chateau Serein (Figure 5.4) is a late, subvertical brittle fault of limited displacement, which displays none of the deformation features to be expected if this were the synmetamorphic, crustal-scale structure inferred by Balé and Brun. Most linear structures within the SBT are subhorizontal or gently plunging and only steepen up into the Cesson and Martin Plage shear zones. The south-side-up sense of movement across these shear zones is clearly incompatible with the proposed south-verging thrust system.

5.6   Post-Cadomian sedimentation

Post-Cadomian red bed deposits rest unconformably on Brioverian sediments in the Plourivo-Plouezec area of the Trégor region (Figure 5.2) and on the Fort de la Latte diorite in the Erquy-Frehel area on the east side of the Baie de St. Brieuc (Figure 5.4). These sediments are commonly correlated with the Rozel Conglomerate Formation of Jersey and the Alderney Sandstone Formation (Figure 5.1) (Renouf, 1974; Roach, 1977). Although all of these units have been variously assigned Devonian or Permian ages as well as

Figure 5.10   Banded hornblendic and quartzo-feldspathic gneisses. Port du Moulin, Sark.

Cambro-Ordovician (Barrois, 1908; Cogné, 1963; Bonhomme *et al.*, 1966), there is now general agreement that they correlate with comparable red beds which occur at or near the base of the Cambrian sequences of Normandy (Doré, 1972). Auvray *et al.* (1980b) obtained an Rb–Sr age of 472 ± 5 Ma from trachy-andesites interbedded within the Plourivo-Plouezec red beds, and trace-fossils within the Cap Frehel red beds are also consistent with an Upper Cambrian–Lower Ordovician age (Bland, B. H., 1984).

The Cap Frehel red beds comprise two stacked fining-upwards cycles of alluvial fan-braided stream type with a minor intercalation of shallow-marine deposits at the top of the first cycle (Went and Andrews, 1990). Sediment dispersal was towards the Cambrian palaeoshoreline which lay to the east in Lower Normandy. Deposition of the red beds of North Brittany and the Channel Islands occurred in basins which were elongate parallel to the ENE–WSW structural grain of the Cadomian belt, and may therefore have formed as a result of the reactivation of Cadomian shear zones and faults.

5.7   Regional correlations

5.7.1   *Guernsey, Sark, Alderney and Cap de la Hague*

Guernsey, Sark, Alderney and Cap de la Hague form the northern margin of the Cadomian belt (Figure 5.1) and despite their mutual geographic separation share a number of geological features in common.

Pre-Cadomian Icartian basement gneisses are well exposed on Guernsey, Sark and Cap de la Hague (Roach 1957; Sutton and Watson, 1957; Adams, 1967, 1976; Leutwein *et al.*, 1973; Power, 1974). The oldest rocks are undated supracrustal sequences of high grade and variably migmatised semi-pelitic schist, quartzite and hornblendic gneiss (Figure 5.10). On Guernsey these are intruded by now highly deformed granitic gneisses, which in turn are intruded by basic sheets, now amphibolites and hornblende schists. Similar granitic gneisses are present on Sark and at Cap de la Hague, but here their relationships to adjacent supracrustal units are less clear. A U–Pb zircon age of 2018 ± 15 Ma obtained from the Icart granite gneiss of Guernsey (Figure 5.11) (Calvez and Vidal, 1978) is generally considered to date syntectonic emplacement of the granitic gneisses during regional high grade metamorphism.

The extent of Icartian basement beneath the Cadomian belt is uncertain. Vidal *et al.* (1981) considered that the low $^{87}Sr/^{86}Sr$ ratios displayed by Cadomian intrusions in the Armorican Massif precluded the presence of extensive areas of continental basement at depth. Accordingly, the NE–SW trending belt of Icartian gneisses present between Cap de la Hague and the Trégor region (Figure 5.11) has been interpreted by some workers (e.g. Rabu *et al.*, 1990) as a fragment of 2000 Ma-old Gondwanan basement detached

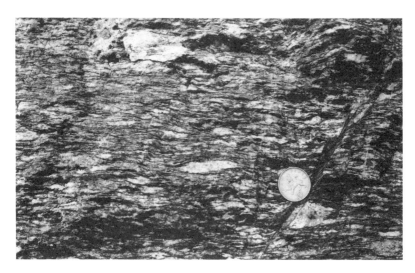

Figure 5.11   Icart granite gneiss, Saint's Bay, Guernsey.

Figure 5.12   Perelle gneiss, Perelle Bay, Guernsey.

from the African craton. Brown *et al.* (1990) have argued, however, that initial $^{87}Sr/^{86}Sr$ ratios do not sufficiently constrain the extent of crustal involvement in magma genesis and suggest that the overall geochemistry of Cadomian intrusions is consistent with the presence of continental basement under much of the Armorican Massif. Sm/Nd studies currently in progress (R. S. D'Lemos, pers. commun. 1990) may help to resolve this problem.

A major suite of early Cadomian foliated plutonic rocks occurs on Guernsey, Sark, Alderney and at Cap de la Hague (Power *et al.*, 1990a) (Figure 5.1). These plutonic rocks, which include the Perelle gneiss of Guernsey (Figure 5.12) and the Thiebot complex of Cap de la Hague, are mainly quartz-dioritic in composition and have a calc-alkaline geochemistry (Power *et al.*, 1990a). Isotopic age data are unsatisfactory but suggest a possible age range of 700–580 Ma. There are no contacts with any rocks of the Brioverian succession. Power *et al.* (1990a) and Strachan *et al.* (1989) have speculated that these foliated granitoids may represent the eroded roots of early Cadomian arcs, possibly analogous to the Penthièvre complex of North Brittany.

The Icart basement gneisses and early Cadomian Perelle gneiss of Guernsey are cut by a swarm of undated dykes termed the Vazon dykes by Lees and Roach (1987). These dykes were metamorphosed and locally heterogeneously deformed during the Cadomian Orogeny, and Lees and Roach (1987) have suggested that the Vazon dykes may be a more northerly expression of the same magmatic cycle that produced the Brioverian Hillion and Erquy volcanics in North Brittany. Low grade Brioverian metasediments are only present east of Cap de la Hague (Power, 1974) and may possibly be represented in southwest Guernsey by the Pleinmont metasediment (Bonney and Hill, 1912).

Late Cadomian post-tectonic, calc-alkaline plutonic complexes occur on Guernsey and Alderney and at Cap de la Hague. The Northern Igneous Complex of Guernsey consists of four main plutonic bodies thought to have been emplaced in the order: St. Peter Port Gabbro, Bordeaux Diorite Complex, Cobo Granite and L'Ancresse Granodiorite (Topley et al., 1990). Field evidence indicates, however, that these bodies were intruded over a short time span since different batches of magma appear to have been contemporaneous. For this reason the 496 ± 13 Ma age of the Cobo Granite (D'Lemos, 1987) probably represents the approximate age of the Northern Igneous Complex as a whole. The Northern Granites Complex of Cap de la Hague has a broadly calc-alkaline character and ranges in composition from diorite to K-feldspar granite (Power et al., 1990b). Rb–Sr whole-rock isochrons indicate that emplacement of the complex spanned c. 50 Ma of Cambrian time in the interval 527–477 Ma (Strachan et al., 1989; Power et al., 1990b) and is thus broadly contemporaneous with similar activity in Guernsey.

5.7.2   Jersey

A 3 km sequence of Brioverian turbidites known as the Jersey 'Shale' Formation crops out in the central and western parts of Jersey. These monotonous sediments have the characteristics of submarine fan deposits (Squire, 1974; Helm and Pickering, 1985). They are overlain by Brioverian andesitic and rhyolitic volcanics in the east and northeast of Jersey. Duff (1978) interpreted an Rb–Sr whole-rock isochron age of 533 ± 16 Ma obtained from the andesites as an extrusion age, implying that deposition of the upper part of the Brioverian volcanic-sedimentary sequence extended into Cambrian times (compare however, Bishop and Mourant, 1979). Both the Jersey Shale Formation and the volcanics have been folded and metamorphosed in the low greenschist facies prior to the emplacement of the post-tectonic granite complexes which form the northwest, southeast and southwest parts of Jersey (Figure 5.1). Precise Rb–Sr whole-rock isochron ages of 550–426 Ma (Bland, A. M., 1984) for the northwest and southwest complexes imply a major phase of Cambrian plutonism in Jersey.

5.7.3   Lower Normandy

The poorly exposed Cadomian rocks of Lower Normandy lie along strike to the northeast of North Brittany (Figure 5.1) and are dominated by variably deformed, low grade Brioverian metasediments and metavolcanics intruded by a variety of calc-alkaline granitoids. Dupret et al. (1990, and references therein) separated the Brioverian into two groups thought to be separated by a major unconformity. Brioverian rocks assigned by them to a lower group are only located north of Granville (Figure 5.1) where they include basic volcanic formations such as Monsurvent and Terrete formations at Coutances, and the La Vast formation east of Cherbourg. These are overlain by black turbiditic mudstones with minor interbedded cherts, volcaniclastics and dolomitic limestones. Syntectonic emplacement of the Coutances quartz diorite at 584 ± 4 Ma (U–Pb zirconage; Guerrot *et al.*, 1986) is thought to have occurred synchronously with uplift of the 'Constantian arc' in the Coutances area.

Subsequent erosion of the 'lower' Brioverian group is thought by Dupret *et al.* (1990) to have produced the turbidites of the 'upper' group which contain fragments and clasts of volcano-plutonic rocks and chert. The 'upper' group mainly crops out south of Granville (Figure 5.1) and was subject to renewed deformation prior to emplacement of the Mancellian batholith at c. 540 Ma. It is emphasised that the contact between the 'lower' and 'upper' Brioverian groups are everywhere faulted and the inferred unconformity is unexposed.

There are a number of points of comparison between the geology of Lower Normandy and North Brittany. The Coutances diorite is of similar age to the Fort de la Latte quartz diorite which lies along strike to the southwest in North Brittany (Figure 5.1), and the Brioverian volcanics of the Coutances area have been compared with the Hillion, Erquy and Fresnaye volcanics which border the Fort de la Latte quartz diorite and Penthièvre complex (Lees *et al.*, 1987; Dupret *et al.*, 1990). The Cadomian rocks south of Granville are the northeastward continuation of the Mancellian Terrane identified in northwest Brittany. The migmatite belts of northeast Brittany do not, however, extend into Lower Normandy. Mylonites southeast of Coutances and a major high strain zone at St. Pair, south of Granville, may represent the along-strike continuation of the Fresnaye and Cancale shear zones, respectively (Figure 5.1), although the kinematics of these structures in Lower Normandy require further investigation.

5.8 Geotectonic setting

Brown *et al.* (1990) have demonstrated that Cadomian magmatism in the North Armorican Massif is calc-alkaline with similarities to high-K orogenic andesites. Intermediate and acid intrusions classify mainly as Volcanic Arc Granites (*sensu* Pearce *et al.*, 1984) irrespective of age or tectonic setting within the North Armorican Massif. We follow Brown *et al.* (1990) in regarding the overall geotectonic setting to be that of a continental margin or Andean arc which was probably located along the northern edge of the Gondwanan Craton during the late Precambrian. Cadomian magmatism and tectono-thermal activity reflects a continuum of active continental margin processes in the hanging wall of a major subduction zone which probably dipped southwards beneath the Gondwanan margin. The precise location of this subduction zone is unclear. It is unlikely to correspond to the English

Channel magnetic anomaly as suggested by Auvray (1979), Peucat et al. (1981), Graviou and Auvray (1985) and Dissler et al. (1988). There is no clear evidence to substantiate a Cadomian age for this anomaly and furthermore the distance between this supposed Cadomian subduction zone and the arc complexes of the Trégor region and their likely correlatives in the Channel Islands and Cap de la Hague (Figure 5.1), is relatively small (c. 50 km) compared with the distances between subduction zones and frontal arcs in contemporary convergent plate settings (> 120 km; Coleman, 1975).

Orogenic activity which may be referred to as 'early' Cadomian is reflected in the formation at c. 700–580 Ma of foliated calc-alkaline complexes such as the Penthièvre complex and possible correlatives in the Channel Islands and Cap de la Hague (Power et al., 1990a). Brioverian volcano-sedimentary sequences accumulated in a variety of marginal basins within the arc complex and were intruded by calc-alkaline plutonic complexes prior to and after their juxtaposition during accretionary tectonism at c. 540 Ma. In this contribution we stress the differences in the Cadomian magmatic and tectono-thermal histories apparent between the three terranes present in North Brittany, each of which represents a displaced segment of the continental margin arc complex. There is insufficient evidence at present to define the original palaeogeographic and relative settings of each of these blocks.

The predominance of sinistral strike-slip displacements across the Cadomian belt in Northern Brittany suggests substantial lateral movements between the components of the belt and implies that orthogonal cross-sections (e.g. Cabanis et al., 1987) are unrealistic. Terrane accretion at c. 540 Ma reflects the oblique collision of crustal blocks in the hanging wall of the subduction zone. The Fresnaye and Cancale shear zones which border the Cadomian terranes of North Brittany may have their analogues along strike to the northeast in Lower Normandy (Figure 5.1) implying a three-terrane accretion model for Brittany and Lower Normandy (Strachan et al., 1989). Although Guernsey, Sark, Alderney and Cap de la Hague have some similarities with parts of the mainland belt, the precise timing and extent of Cadomian activity here is poorly constrained and this northern margin of the Armorican Massif may constitute an additional terrane (cf. Power et al., 1990a).

Our interpretation that the Cadomian belt in Brittany comprises a series of displaced terranes which acquired their main characteristics prior to amalgamation along strike-slip shear zones contrasts with the collision model proposed by Graviou and Auvray (1985) and Graviou et al. (1988). These workers argue that crustal thickening in response to regional southward-directed thrusting (cf. Balé and Brun, 1983) resulted in crustal anatexis and the post-tectonic formation of the migmatite belts and the granites of the Mancellian batholith. We agree with Brown et al. (1990) that this model is inconsistent with both the field evidence and the geochemical characteristics of the Mancellian granites, and unlikely given the present average crustal thickness (Bois and Courtillot, 1988) in combination with only low metamorphic grade at the present erosion level. The deposition of substantial quantities of Cambrian sediment of Normandy at or near sea level, within the orogenic belt, further implies that no major crustal thickening took place during the Cadomian in this region.

Acknowledgements

The authors acknowledge research funding and technical support from Oxford Polytechnic (R.A.S., P.J.T.) and the University of Keele (R.A.R.), and discussions with R. S. D'Lemos, R. D. Dallmeyer, M. Brown and G. Power.

References

Adams, C. J. D. (1967) A geochronological and related isotopic study of rocks from northwest France and the Channel Islands (United Kingdom). Ph.D. thesis, University of Oxford.
Adams, C. J. D. (1976) Geochronology of the Channel Islands and the adjacent French mainland. *J. Geol. Soc. London* **132**, 233–250.
Andriamarofohatra, J. and La Boisse, H. (1988) Mise en évidence de termoins granitiques tardicadomiens à 540 Ma dans la region de Belle Isle en Terre, Massif Armoricain. *Bull. Soc. Geol. France* **8**, 279–287.
Autran, A., Chantraine, J. and Rabu, D. (1979) Lithostratigraphie et deformation du Brioverien de la baie de Lannion. Implications sur les relations entre les cycles cadomien et hercynien. *Bull. BRGM Orleans* (2), I, **4**, 277–292.
Auvray, B. (1979) Genèse et évolution de la crôute continentale dans le Nord du Massif armoricain. Thèse Etat, Rennes.
Auvray, B., Monnier, J. L. and Lefort, J. P. (1976) *Carte Geologique de la France a 1:50,000, No 171.* 1st edn. (avec notice explicative 26 pp).
Auvray, B., Charlot, R. and Vidal, P. (1980a) Données nouvelles sur le Protérozoique inférieur du domaine nord-Armoricain (France): âge et signification. *Can. J. Earth Sci* **17**, 532–538.
Auvray, B., Mace, J., Vidal, P. and Van der Voo, R. (1980b) Rb–Sr dating of the Plouezec volcanics, northern Brittany: implications for the age of red-bed (series rouges) in the northern Armorican Massif. *J. Geol. Soc. London* **137**, 207–210.
Balé, P. (1986) Tectonique cadomienne en Bretagne nord. Thèse d'Université, Rennes.
Balé, P. and Brun, J. P. (1983) Les chevauchments cadomiens dans le baie de Saint-Brieuc. *C.R. Acad. Sci., Paris* **297**, 359–362.
Balé, P. and Brun, J. P. (1989) Late Precambrian thrust and wrench zones in northern Brittany. *J. Struct. Geol.* **11**, 391–406.
Barrois, C. (1895) Sur les poudingues de Cesson. *Ann. Soc. Geol. Nord* **23**, 26–29.
Barrois, C. (1896) Legende de la feuille de Saint-Brieuc (No 59 de la Carte geologique de France au 1:80:000). *Ann. Soc. Geol. Nord* **23**, 66–87.
Barrois, C. (1908) Legende de la feuille de Tréguier (No 42 de la carte geologique de France au 1:80,000). *Ann. Soc. Geol. Nord* **23**, 66–87.
Bertrand, L. (1921) *Les Anciennes Mers de la France et leurs Depots*. Flammarion, Paris.
Bishop, A. C. and Mourant, A. E. (1979) Discussion on the Rb–Sr whole-rock age determination of the Jersey Andesite Formation, Jersey, C.I. *J. Geol. Soc. London* **136**, 121–122.
Bland, A. M. (1984) The geology of the granites of western Jersey, with particular reference to the Southwest Granite complex. Ph.D. thesis, CNAA, (Oxford Polytechnic).
Bland, B. H. (1984) *Arumberia* Glaessner and Walter, a review of its potential for correlation in the region of the Precambrian–Cambrian boundary. *Geol. Mag.* **121**, 625–633.
Bois, C. and Courtillot, V. (1988) The French ECORS Program. Deep Seismic Profiling of the Crust and Evolution of the Lithosphere. *EOS* **25** October 1988, 977 and 989.
Bond, S. G., Nickerson, P. A. and Kominz, M. A. (1984) Breakup of a supercontinent between

625 Ma and 555 Ma: New evidence and implications for continental histories. *Earth Planet. Sci. Lett.* **70**, 325–345.

Bonhomme, M., Cogné, J., Leutwein, F. and Sonnet, J. (1966) Données nouvelles sur l'âge des series rouges du golfe normanno-breton *C.R. Acad. Sci., Paris* **262**, 606–609.

Bonney, T. G. and Hill, E. (1912) Petrological notes on Guernsey, Herm, Sark and Alderney. *Q. J. Geol. Soc. London* **68**, 31–67.

Brown, M. (1978) The tectonic evolution of the Precambrian rocks of the St. Malo region, Armorican Massif, France. *Precambrian Res.* **6**, 1–21.

Brown, M. (1979) The Petrogenesis of the St. Malo Migmatite Belt, Armorican Massif, France, with particular reference to the diatexites. *Neues Jahrb. Mineral., Abh.* **135**, 48–74.

Brown, M., Power, G. M., Topley, C. G. and D'Lemos, R. S. (1990) Cadomian magmatism in the North Armorican Massif. In *The Cadomian Orogeny*, eds. D'Lemos, R. S., Strachan, R. A. and Topley, C. G., *Spec. Publ. Geol. Soc. London* **51**, 181–213.

Brun, J. P. (1977) La zonation structurale des domes gneissique. Un exemple le Massif de St. Malo (Massif Armoricain, France). *Can. J. Earth Sci.* **14**, 1697–1707.

Brun, J. P. and Balé, P. (1990) Cadomian tectonics in Northern Brittany. In *The Cadomian Orogeny*, eds. D'Lemos, R. S., Strachan, R. A. and Topley, C. G., *Spec. Publ. Geol. Soc. London* **51**, 95–114.

Brun, J. P. and Martin, H. (1979) The St. Malo migmatite belt: a late Precambrian gneiss dome: a comment. *Precambrian Res.* 137–143.

Cabanis, B., Chantraine, J. and Rabu, D. (1987) Geochemical study of the Brioverian (Late Proterozoic) volcanic rocks in the northern Armorican Massif (France). In *Geochemistry and Mineralisation of Proterozoic Volcanic Suites*, eds. Pharoah, T., Beckinsale, R. D. and Rickard, D. T., *Spec. Publ. Geol. Soc. London* **33**, 525–539.

Calvez, J. Y. and Vidal, P. (1978) Two billion years old relics in the Hercynian belt of Western Europe. *Contrib. Mineral. Petrol.* **65**, 395–399.

Chantraine, J., Chauvel, J. J., Dupret, L., Gatinot, F., Icart, J. C., Le Corre, C., Rabu, D., Sauvan, P. and Villey, M. (1982) Inventaire lithologique et structural du Briovérien (Protérozoique supérieur) de la Bretagne centrale et du Bocage normand. *Bull. BRGM* Orleans (2), **1–2**, 3–18.

Chantraine, J., Chauvel, J. J., Bale, P., Denis, E. and Rabu, D. (1988) Le Briovérien (Protérozoique supérieur a terminal) et l'orogenese cadomienne en Bretagne (France). *Bull. Soc. Geol. France* (8), **IV**, 5.

Cobbold, P. R. and Quinquis, H. (1980) Development of sheath folds in shear regimes. *J. Struct. Geol.* **2**, 119–126.

Cogné, J. (1959) Données nouvelles sur l'Antecambrien dans l'ouest de la France: Pentevrién et Brioverién en baie de Saint-Brieuc (Côtes-du-Nord). *Bull. Soc. Geol. France* (7), **I**, 112–118.

Cogné, J. (1962) Le Briovérien: esquisse des caracteres stratigraphiques, metamorphiques, structuraux et paleogeographiques de l'Antecambrien recent dans le Massif armoricain. *Bull. Soc. Geol. France* (7), **IX**, 413–430.

Cogné, J. (1963) Reflexion sur l'age des series detritiques rouges du Nord de l'Armorique (golfe normanno-breton). *Bull. Soc. Geol. Min. Bret.* 17–30.

Cogné, J. and Wright, A. E. (1980) L'orogene cadomien. Vers un essai d'interpretations paleogeo-dynamique unitaire des phenomenes orogeniques fini-precambriens d'Europe moyenne et occidentale, et leur signification a l'origine de la croûte et du mobilisme varisque puis alpin. In *Geologie de l'Europe du Precambrien aux Bassins Sedimentaires Post-Hercyniens*, eds. Cogné, J. and Slansky, M., 26ᵉ *Congres Geologique Internationale, Colloque C6*, Paris.

Coleman, P. J. (1975) On island arcs. *Earth-Sci. Rev.* **11**, 47–80.

Denis, E. and Dabard, M. P. (1988) Sandstone petrography and geochemistry of late Proterozoic sediments of the Armorican Massif (France)—a key to basin evolution during the Cadomian Orogeny. *Precambrian Res.* **42**, 189–206.

Dissler, E., Doré, F., Dupret, L., Gresselin, F. and Le Gall, J. (1988) L'evolution géodynamique cadomienne du Nord-Est du Massif armoricain. *Bull. Soc. Geol. France* (8), **IV**, 5, 801–814.

D'Lemos, R. S. (1987) The evolution of the Northern Igneous Complex of Guernsey, Channel Islands—some isotopic evidence. *Proc. Ussher Soc.* **6**, 498–501.

Doré, F. (1972) La transgression majeure du Palaeozique inferieur dans le Nord-Est du massif Armoricain. *Bull. Geol. Soc. France* (7), **14**, 79–93.

Duff, B. A. (1978) Rb–Sr whole-rock age determination on the Jersey Andesite formation, Jersey, C.I. *J. Geol. Soc. London* **135**, 153–156.

Dupret, L., Dissler, E., Doré, F., Gresselein, F. and Le Gall, J. (1990) Cadomian geodynamic evolution of the Northeastern Armorican Massif (Normandy and Maine). In *The Cadomian Orogeny*, eds. D'Lemos, R. S., Strachan, R. A. and Topley, C. G., *Spec. Publ. Geol. Soc, London* **51**, 115–131.

Gapais, D. and Le Corre, C. (1980) Is the Hercynian belt of Brittany a major shear zone? *Nature (London)* **288**, 574–576.

Graindor, M. J. (1957) Le Brioverién dans le Nord-Est du Massif armoricain. Memoires pour servir a l'explication de la Carte Geologique detaillée de la France, BRGM.

Graviou, P. and Auvray, B. (1985) Caractérisation pétrographique et géochimique des granitoides cadomiens du domaine nord-armoricain: implications géodynamiques. *C.R. Acad. Sci., Paris* **301**, 1315–1318.

Graviou, P., Peucat, J. J., Auvray, B. and Vidal, P. (1988) The Cadomian orogeny in the northern Armorican Massif: petrological and geochronological constraints on a geodynamic model. *Hercynica*, **4**, 1–13.

Griffiths, N. H. (1985) The geology of the Morlaix—St Michel-en-Greve region, NW Brittany, France. Unpublished PhD thesis, University of Keele.

Guerrot, C. and Peucat, J. J. (1990) U–Pb geochronology of the Late Proterozoic Cadomian Orogeny in the Northern Armorican Massif, France. In *The Cadomian Orogeny*, eds. D'Lemos, R. S., Strachan, R. A. and Topley, C. G., *Spec. Publ. Geol. Soc. London* **51**, 29–42.

Guerrot, C., Peucat, J. J. and Dupret, L. (1986) Age du Precambrien sedimentaire (Brioverién) dans le Massif armoricain. *lle Reun. Sci. Terre*, Clermont-Ferrand.

Hanmer, S., Le Corre, C. and Berthé, D. (1982) The role of Hercynian granites in the deformation and metamorphism of Brioverian and Palaeozoic rocks of Central Brittany. *J. Geol. Soc. London* **139**, 85–94.

Harding, T. P. (1985) Seismic characteristics and identification of negative flower structures, positive flower structures and positive structural inversion. *Am. Assoc. Pet. Geol. Bull.* **69**, 582–600.

Helm, D. G. and Pickering, K. T. (1985) The Jersey Shale Formation. A late Precambrian deep-water siliciclastic system, Jersey, Channel Islands. *Sediment. Geol.* **43**, 43–66.

Jeanette, D. (1971) Analyse Tectonique de Formations précambriennes. Etude du Nord-Est de la Bretagne. Thèse d'Université, Strasbourg.

Jeanette, D. and Cogné, J. (1968) Une discordance majeure au sein du Brioverién au flanc ouest de la baie de Saint-Brieuc. *C.R. Acad. Sci. Paris* **226**, 2211–2214.

Jonin, M. (1973) Les differents types granitiques de la Mancellian et l'unite du batholite maneau (Massif armoricain). *C.R. Acad. Sci. Paris* **256**, 2006.

Jonin, M. (1981) Un batholite fini-Precambrien: le Batholite Mancellien (Massif armoricain). Thèse, Docteur-es-Sciences Naturelles, Université de Bretagne Occidentale (Brest), 320 pp.

Jonin, M. and Vidal, P. (1975) Etude geochronologique des granitoides de la Mancellia, Massif Armoricain, France. *Can. J. Earth Sci.* **12**, 920–927.

Le Corre, C. (1977) Le Brioverién de Bretagne centrale. *Bull. BRGM Orleans* **1–3**, 219–254.

Lees, G. J. and Roach, R. A. (1987) The Vazon Dyke Swarm, Guernsey, Channel Islands. *Proc. Ussher Soc.* **6**, 502–509.

Lees, G. J., Roach, R. A., Shufflebotham, M. M. and Griffiths, N. H. (1987) Upper Proterozoic basaltic Volcanism in the northern Massif Armoricain, France. In *Geochemistry and Mineralisation of Proterozoic Volcanic Suites*, eds. Pharaoh, T. C., Beckinsale, R. D. and Rickard, D. T., *Spec. Publ. Geol. Soc. London* **33**, 503–525.

Le Gall, J. and Mary, G. (1982) Mise en place tardi-cadomienne du complexe basique de Bree (Mayenne). *Bull. BRGM Orleans* **1–2**, 19–23.

Le Gall, J. and Mary, G. (1983) Misé en place tardi-cadomienne du complexe basique de Bree et des autres venues gabbroiques et doleritiques dans l'histoire cadomo-varisque de l'Est du Massif armoricain. *Bull. Soc. Geol. Min. Bret.* **15**, 169–180.

Leutwein, F., Sonet, J. and Zimmermann, J. L. (1968) Géochronologie et evolution orogenique précambrienne et hercynienne de la partie nord-est du Massif Armoricain. *Sciences de la Terre, Mem.* **11**, Nancy.

Leutwein, F., Power, G. M., Roach, R. A. and Sonet, J. (1973) Quelques résultats obtenus sur des roches d'âge précambrien du Cotentin. *C.R. Acad. Sci. Paris* **276**, 2121–2124.

Pearce, J. A., Harris, N. B. W. and Tindle, A. G. (1984) Trace element discrimination diagrams for the tectonic interpretation of granitic rocks. *J. Petrol.* **25**, 956–983.

Peucat, J. J. (1986) Behaviour of Rb–Sr whole rock and U–Pb zircon systems during partial melting as shown in migmatitic gneisses from the St. Malo Massif, NE Brittany, France. *J. Geol. Soc. London*, **143**, 875–886.

Peucat, J. J., Hirbec, Y., Auvray, B., Cogné, J. and Cornichet, J. (1981) Late Proterozoic zircon age from a basic–ultrabasic complex: a possible Cadomian orogenic complex in the Hercynian belt of Western Europe. *Geology*, 169–173.

Power, G. M. (1974) The geology of the Precambrian rocks of La Hague, Manche, France. PhD thesis, University of Keele.

Power, G. M., Brewer, T. S., Brown, M. and Gibbons, W. (1990a) Late Precambrian foliated plutonic complexes of the Channel Islands and La Hague—Early Cadomian Plutonism. In *The Cadomian Orogeny*, eds. D'Lemos, R. S., Strachan, R. A. and Topley, C. G., *Spec. Publ. Geol. Soc. London* **51**, 57–71.

Power, G. M., Brewer, T. S. and D'Lemos, R. S. (1990b) The post-tectonic Cadomian plutonic complex of La Hague, Manche, France. In *The Cadomian Orogeny*, eds. D'Lemos, R. S., Strachan, R. A. and Topley, C. G., *Spec. Publ. Geol. Soc. London* **51**, 261–272.

Rabu, D., Chauvel, J. J. and Chantraine, J. (1982) Le domaine interne de la chaîne cadomienne dans le Massif Armoricain étude lithostratigraphique, géochemique et structurale le long d'un traversale en baie de Saint-Brieuc. *Rapp. BRGM Document BRGM, No 66*, (1983) Orleans 31 pp.

Rabu, D., Chauvel, J. J. and Chantraine, J. (1983) Nouvelles propositions pour la lithostratigraphie du Brioverién (Protérozoïque supérieur) et pour l'évolution géodynamique cadomienne en baie de Saint-Brieuc. (massif Armoricain). *Bull. Soc. Geol. France* (7), **XXV**, 615–621.

Rabu, D., Chantraine, J., Chauvel, J. J., Denis, E., Balé, P. and Bardy, P. H. (1990) The Brioverian (Late Proterozoic) and the Cadomian orogeny in the Armorican Massif. In *The Cadomian Orogeny*, eds. D'Lemos, R. S., Strachan, R. A. and Topley, C. G., *Spec. Publ. Geol. Soc. London* (in press).

Renouf, J. T. (1974) The Proterozoic and Palaeozoic development of the Armorican and Cornubian provinces. *Proc. Ussher Soc.* **3**, 6–43.

Roach, R. A. (1957) The geology of the metamorphic complex of south and central Guernsey. PhD thesis, University of Nottingham.

Roach, R. A. (1977) A review of the Precambrian rocks of the British Variscides and their relationships with the Precambrian of northwest France. In La Chaine Varisque d'Europe Moyenne et Occidentale. *Coll Int. CNRS, Rennes* **243**, 61–79.

Roach, R. A., Adams, C., Brown, M., Power, G. M. and Ryan, P. D. (1972) The Precambrian stratigraphy of the Armorican Massif, northwest France. *24th Int. Geol. Congr. Montreal* **1**, 246–252.

Roach, R. A., Topley, C. G., Brown, M. and Shufflebotham, M. (1986) Brioverian volcanism and Cadomian plutonism in the northern part of the Armorican Massif. *Pre-Conference Excursion Guide to 1GCP 217 Meeting*, Keyworth, Notts., UK, 1–5 April 1986, 90 pp.

Roach, R. A., Strachan, R. A., Shufflebotham, M., Power, G. M., Dupret, L. and Brown, M. (1988) The Cadomian belt of Northern Brittany and Lower Normandy: a field guide. *Post-Conference Excursion Guide to IGCP 233 Meeting*, Oxford Polytechnic, Oxford, UK, 6–8 April 1988, 112 pp.

Roach, R. A., Lees, G. J. and Shufflebotham, M. (1990) Basaltic volcanism in the Baie de St Brieuc, North Brittany: Early stages in the development of a Late Proterozoic ensialic basin. In *The Cadomian Orogeny*, eds. D'Lemos, R. S., Strachan, R. A. and Topley, C. G., *Spec. Publ. Geol. Soc. London* **51**, 41–67.

Ryan, P. D. (1973) The solid geology of the area between Binic and Brehec, Cotes-du-Nord, France. Unpublished PhD thesis, University of Keele.

Ryan, P. D. and Roach, R. A. (1975) The Brioverian-Pentevrian boundary at Palus Plage (Armoricain Massif, France). *Bull. Soc. Geol. Min. Bret.* **7**, 1–20.

Shufflebotham, M. (1989) Geochemistry and geotectonic interpretation of the Penthièvre crystalline massif, N. Brittany, France. *Precambrian Res.* **45**, 247–261.

Shufflebotham, M. (1990) The geology of the Penthièvre crystalline massif: a reappraisal of the type-Pentevrian area, Northern Brittany. In *The Cadomian Orogeny*, eds. D'Lemos, R. S., Strachan, R. A. and Topley, C. G., *Spec. Publ. Geol. Soc. London* **51**, 43–55.

Squire, A. D. (1974) Brioverian sedimentology and structure of Jersey and adjacent areas. PhD thesis, University of London.

Strachan, R. A. and Roach, R. A. (1990) Tectonic evolution of the Cadomian belt in North

Brittany. In *The Cadomian Orogeny*, eds. D'Lemos, R. S., Strachan, R. A. and Topley, C. G., *Spec. Publ. Geol. Soc. London* **51**, 133–150.

Strachan, R. A., Treloar, P. J., Brown, M. and D'Lemos, R. S. (1989) Cadomian terrane tectonics and magmatism in the Armorican Massif. *J. Geol. Soc. London* **46**, 423–426.

Sutton, J. and Watson, J. V. (1957) The structure of Sark, Channel Islands. *Proc. Geol. Assoc. London* **68**, 179–203.

Topley, C. G., Brown, M., Power, G. M. and D'Lemos, R. S. (1990) The Northern Igneous Complex of Guernsey. In *The Cadomian Orogeny*, eds. D'Lemos, R. S., Strachan, R. A. and Topley, C. G., *Spec. Publ. Geol. Soc. London* **51**, 245–261.

Treloar, P. J. and Strachan, R. A. (1990) Cadomian strike-slip tectonics in northeast Brittany. In *The Cadomian Orogeny*, eds. D'Lemos, R. S., Strachan, R. A. and Topley, C. G., *Spec. Publ. Geol. Soc. London* **51**, 151–168.

Verdier, P. (1968) Etude petrographique et structurale de Tregor ocidentale (Baie de Lannion: Cotes-du-Nord, Finistere). These 3eme cycle, Universite de Strasbourg.

Vidal, P. (1980) L'evolution du Massif Armoricain: apport de la géochronologie et de la géochemie isotopique du strontium. *Mem. Soc. Geol. Min. Bret.* **21**.

Vidal, P., Auvray, B., Chauvet, J. F. and Cogné, J. (1972) L'âge radiométrique de la diorite de St Quay-Portrieux (Côtes-du-Nord). Ses conséquences sur le Broverién de la baie de Saint-Brieuc. *C.R. Acad. Sci., Paris* **275**, 1323–1326.

Vidal, P., Deutsch, S., Martineau, F. and Cogné, J. (1974) Nouvelles données radiometriques en baie de Saint-Brieuc. Le probleme d'un socle ante-cadomien nord-armoricain. *C.R. Acad. Sci., Paris* **279**, 631–634.

Vidal, P., Auvray, B., Charlot, R. and Cogné, J. (1981) Precadomian relicts in the Armorican Massif: their age and role in the evolution of the western and central European Cadomian-Hercynian belt. *Precambrian Res.* **14**, 1–20.

Watts, M. J. and Williams, G. D. (1979) Fault rocks as indicators of progressive shear deformation in the Guingamp region, Brittany. *J. Struct. Geol.* **1**, 323–332.

Went, D. and Andrews, M. (1990) Post-Cadomian erosion, deposition and basin development in the Channel Islands and Northern Brittany. In *The Cadomian Orogeny*, eds. D'Lemos, R. S., Strachan, R. A. and Topley, C. G., *Spec. Publ. Geol. Soc. London* **51**, 293–304.

White, A. J. R. and Chappell, B. W. (1983) Granitoid types and their distribution in the Lachlan Fold Belt, southeastern Australia. *Mem. Geol. Soc. Am.* **159**, 21–34.

6 Precambrian of the Bohemian Massif, Central Europe

J. CHALOUPSKÝ

6.1 Introduction

The extent and nature of Precambrian events within the Bohemian Massif are controversial. The same metamorphic rock units have variously been assigned either to the Lower Palaeozoic or the Proterozoic, and structures and metamorphic phenomena which affect these rocks have accordingly been designated as Variscan, Caledonian, Cadomian or even pre-Cadomian. In recent years there has been a tendency to consider that all the massif comprises rocks of Upper Proterozoic to Lower Palaeozoic age, with the main structural and metamorphic features of Variscan age. This paper suggests that the extent of Precambrian events in the massif has been underestimated. It presents a new stratigraphic subdivision of the Precambrian units and summarises their tectonic and metamorphic evolution.

6.2 The Bohemian Massif

The Bohemian Massif represents the largest exposed part of the Variscan (Hercynian) orogenic belt in Central Europe (Figure 6.1A). It is mainly exposed in Czechoslovakia, the marginal parts extending to Austria, the Federal Republic of Germany, the German Democratic Republic and Poland. In the south and southeast the massif plunges beneath the East Alpine-Carpathian Tertiary Foredeep and nappes.

Within the massif, folded and regionally metamorphosed Precambrian and Lower Palaeozoic units are unconformably overlain by Permo-Carboniferous molasse sediments and Upper Cretaceous to Tertiary platform successions. These overstep sequences are widespread in northern Bohemia, where they attain thicknesses of up to 2000 m.

During the Precambrian and Lower Palaeozoic, periods of rifting alternated with compressive orogenic events of differing intensities and extent (pre-Cadomian, Cadomian, Caledonian and Variscan). The result is a collage

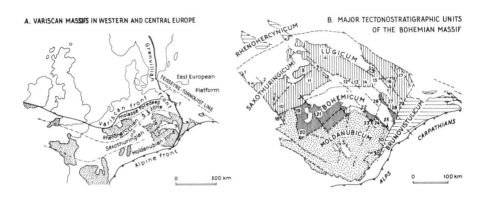

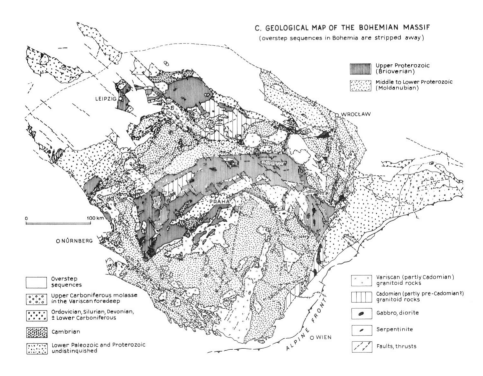

Figure 6.1   Precambrian and Lower Palaeozoic units of the Bohemian Massif. Numbers in **B** indicate the position of significant areas or localities mentioned in the text: 1 Central German Crystalline Rise; 2 Schwarzburg; 3 Leipzig; 4 Lusatia; 5 Kaczawa Mts; 6 Munchberg Gneiss Massif; 7 Wildenfels, Central Saxonian Lineament; 8 Saxon Granulite Massif; 9 Elbe/Labe zone; 10 Fichtelgebirge/Smrciny Mts; 11 Erzgebirge/Krusne hory Mts; 12 Jested Mts; 13 Zelezny Brod; 14 Krkonose Mts; 15 Leszcyniec; 16 Sowie Gory Gneiss Massif; 17 Fore-Sudentic block; 18 Erbendorf; 19 Marianske Lazne; 20 Kdyne; 21 Tepla-Barrandian area; 22 Zelezne hory Mts; 23 Ransko; 24 Vitanov; 25 Letovice; 26 Nove Mesto; 27 Orlicke hory-Klodzko; 28 Stare Mesto; 29 Jeseniky Mts Silesicum; 30 Moravicum.

of tectonostratigraphic units which differ in their geological evolution and structure. Five major tectonostratigraphic units, each about 100–200 km wide, have been delimited in the massif. These are the Moldanubicum, Bohemicum and Saxothuringicum-Lugicum in the inner part of the massif, and the Brunovistulicum and Rhenohercynicum in its outer part (Figure 6.1B). The units are mostly bounded by deep-seated faults, thrusts and mylonite zones and may be interpreted as terranes. They may alternatively represent para-autochthonous units whose lateral movements were restricted and whose contrasting Upper Proterozoic and Palaeozoic geology results from pre-existing inhomogeneities of the Precambrian continental crust (Chaloupsky, 1988).

The Lower Palaeozoic basins developed entirely in an ensialic environment. They differ in their degree of subsidence, thickness of sediment and in the timing of the initiation and cessation of sedimentation. The evolution of the basins as well as the style of the Variscan and restricted Caledonian deformations were largely controlled by the character and tectonic structures of the Cadomian or pre-Cadomian basement. A rejuvenation of the Lower Palaeozoic basins and domination of the Caledonian to Variscan orogenic phases from the central to the outer parts of the massif is well documented. Most of the Lower Palaeozoic rocks are unmetamorphosed or only underwent low grade metamorphism. Intense deformation took place especially along the tectonic boundaries of the Precambrian blocks. An extensive phase of Carboniferous plutonism marks the final Variscan tectono-thermal event.

The Precambrian of the Bohemian Massif can be divided into two major units which differ in lithology, structure and degree of metamorphism. The first is of Upper Proterozoic age and consists of folded, low- to medium-grade metasediments. These are lithologically monotonous greywackes, with local basic volcanics. The unit mainly occurs in the Tepla-Barrandian area southwest of Prague (Figure 6.1C). It has often been referred to as 'Algonkian' (Kettner, 1917) and correlated with the Brioverian of the Armorican Massif in France (Zoubek, 1988a).

The second unit is generally more intensely metamorphosed and includes micaschists, gneisses and migmatites with common intercalations of crystalline limestone, quartzite, graphite schist, calc-silicate rock and amphibolite. This unit is mainly exposed in the Moldanubicum area of southern Bohemia and has been often considered to be Middle to Lower Proterozoic in age.

In other parts of the Bohemian Massif the Precambrian rocks have a more limited extent and are separated by and largely covered by Palaeozoic and younger successions. Most of them have been correlated with the Upper Proterozoic sequences of the Tepla-Barrandian area, with only the high grade metamorphic rocks in the Sowie Gory Gneiss Massif in Poland and the Saxon Granulite Massif in the GDR compared with Moldanubicum rocks

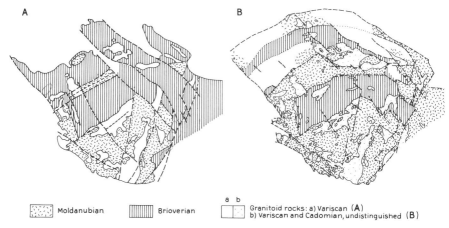

A   B

| Moldanubian | Brioverian | Granitoid rocks: a) Variscan (**A**) b) Variscan and Cadomian, undistinguished (**B**) |

Figure 6.2   Distribution of Precambrian units in the Bohemian Massif (Palaeozoic and younger sequences are stripped away). **A.** after Zoubek (1988); **B.** present paper.

(Kodym, 1946; Maska *et al.*, 1960; Hoth *et al.*, 1979; Zoubek, 1988a) (Figure 6.2A). The extent of the Moldanubian rocks is much greater than previously assumed (Chaloupsky, 1988) (Figure 6.2A).

The Upper Proterozoic and pre-Upper Proterozoic units of the Bohemian Massif have been recently delimited and classified as regional chronostratigraphic units and termed, respectively, the Brioverian and Moldanubian.

Not all geologists regard the Moldanubian and Brioverian as superposed stratigraphic units but rather as diverse components of a single Upper Proterozoic complex (e.g. Vejnar, 1965; Pertold and Suk, 1986; Stettner, 1988). Some geologists have gone still further and classified the Moldanubian rocks as Lower Palaeozoic, at least in part. They have stressed the strong predominance of Variscan radiometric ages in the Moldanubian rocks, as well as their lithological similarities to fossiliferous Lower Palaeozoic formations and the presence of microfossils similar to Palaeozoic forms (e.g. Konzalova, 1988; Wojciechowska and Gunia, 1988; Reitz *et al.* 1988). However, the observation that metamorphic zones pass from the Brioverian into the Moldanubian units without break does not necessarily mean that the units are of similar age; this could simply indicate overprinting by similar Cadomian and/or Variscan tectono-thermal processes. Abundant Variscan radiometric ages in the Precambrian metamorphic rocks of the massif clearly mark the last major tectono-thermal event and overprint. The rare microfossils in the Moldanubian are either problematic with regard to their age, or are found in rocks which need not be an integral part of the Moldanubian complex (Konzalova, 1988; Zoubek, 1988a). The Moldanubian rocks always form the core of major anticlinal structures of the massif and the Brioverian rocks the envelope. The Moldanubian rocks are therefore likely to be older.

6.3   The Moldanubian

The Moldanubian includes all Precambrian rocks older than the Upper Proterozoic (Brioverian) sequences of the Bohemicum. The Moldanubian complex consists mostly of medium- to high-grade metamorphic rocks. It comprises the whole Moldanubicum area of south Bohemia, several minor blocks in the Saxothuringicum-Lugicum, and is the basement to the thick Devonian and Carboniferous successions in the Brunovistulicum.

In the Moldanubicum area the crystalline complex has been subdivided by Zoubek (1988a) into four groups:

(1) The Zeliv (Monotonous) group (> 3000 m thick) composed of paragneisses and migmatites, which include at higher levels scarce thin intercalations of quartzite, calc-silicate rocks and metabasites.

(2) The Klet (Leptynitic) group (50–2000 m thick) consisting mostly of granulite rocks interlayered with amphibolites.

(3) The varied group (100–1000 m thick) composed of paragneisses and mica schists rich in intercalations of crystalline limestone, graphite schist, quartzite, calc-silicate rock and amphibolite.

(4) The Kaplice (Flysch) group (> 2000 m thick) composed of mica schists and paragneisses, with local quartzites (Figure 6.3).

The Klet group granulites have been regarded either as metamorphosed volcano-sedimentary units, which occupy a distinct stratigraphic position within the Moldanubian complex (Jencek and Vajner, 1968; Suk, 1974;

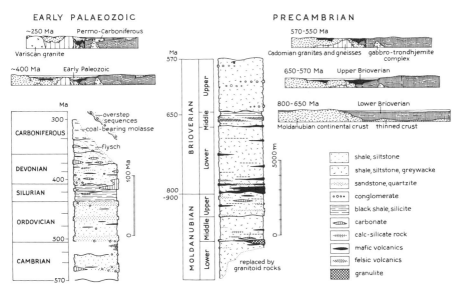

Figure 6.3   Generalised stratigraphic columns and proposed late Proterozoic to early Palaeozoic tectono-thermal evolution in the Bohemian Massif.

Zoubek, 1988a) or as allochthonous bodies tectonically transported upwards from deeper crustal levels (Vrana, 1979; Behr et al., 1984; Zeman, 1984; Fiala et al., 1987). The Moldanubian granulites typically occur in association with amphibolites and ultramafic rocks such as peridotites, serpentinites and eclogites (Fediukova and Dudek, 1977; Machart, 1984).

In the Saxothuringicum and Lugicum areas the higher, lithologically more varied units predominate. They occur in the western Erzgebirge and Fichtelgebirge Mts, the Krkonose Mts, the Orlicke hory–Klodzko unit, the Fore-Sudetic Block and the Jeseniky Mts. The lower, more monotonous units with local granulites are restricted mainly to the elevated or upthrusted and deeply denudated cores of the Saxon Granulite Massif, the eastern Erzgebirge/Krusne hory Mts, the Grossenhain block in the Elbe/Labe zone and the Sowie Gory Gneiss Massif.

In the Brunovistulicum area the Precambrian basement is composed of plutonic rocks and crystalline schists whose stratigraphic affinities are unknown. Gneisses and migmatites found in boreholes as well as the exposed Desna gneiss in the Silesicum may represent some stratigraphic equivalents of the oldest monotonous group of the Moldanubicum complex. Units comparable with the Moldanubian Varied Group have not been identified in this area (Dudek, 1980).

The common components of the Moldanubian complex are granite-gneisses, orthogneisses and migmatites which form thick concordant bodies in the lower portions of the Varied Group. They are mostly interpreted as the products of granitisation of Moldanubian supracrustal rocks in the Pre-cambrian orogenies. The Varied Group is also characterised by magnetite and polymetallic skarns. They belong to the oldest, pre-Upper Proterozoic period of mineralization in the Bohemian Massif, and are possibly of volcano-sedimentary origin.

There is one Precambrian unit in the Bohemian Massif whose stratigraphic affinities are unknown. It comprises phyllites, mica schists, quartzites, with metabasites and carbonates. It has been various assigned to the Ordovician, Cambrian and/or Upper Proterozoic. It is exposed in the Moravicum, a significant tectonic zone between the Moldanubicum and Brunovistulicum, and on the flanks of the Erzgebirge/Krusne hory–Fichtelgebirge/Smrciny anticlinal zone in the Saxothuringicum. It may be of similar age to some quartzite-phyllite-mica schist complexes which occur in the Brunovistulicum basement, in the Fore/sudetic Block and the Central German Crystalline Rise. In the Moravicum the complex is about 1500 m thick and consists of two groups separated by the stratiform Bites gneiss: the lower, Bily Potok Group containing quartzite intercalations and the upper, Olesnice Group which also includes crystalline limestone, graphite schist and metabasite (Jaros and Misar, 1976; Zoubek, 1988a).

In the Erzgebirge/Krusne hory and Fichtelgebirge/Smrciny Mts the phyllite–mica schist complex overlies units correlated with the Moldanubian

Varied Group. It is unconformably overlain by Cambrian to Lower Ordovician sediments. Stratigraphically the complex occupies the uppermost position in the Moldanubian complex (Figure 6.3). It might be correlated with very similar c. 1250–1000 Ma-old Dalslandian units in the southwestern part of the Baltic Shield. Correlation of the Moravicum units with the Moldamibian varied group is not excluded.

For future lithostratigraphic correlations the following tripartite subdivision of the pre-Upper Proterozoic Moldanubian complex of the Bohemian Massif is most convenient (Figure 6.3).

(1) The Lower Moldanubian, including the monotonous group of paragneisses, migmatites and granulites.

(2) The Middle Moldanubian, including the Varied Group overlain by a more monotonous sequence of mica schists.

(3) The Upper Moldanubian, including the quartzite-phyllite-mica schist complex, with local metabasites.

6.4 The Brioverian

The Brioverian sequences comprise the whole of the Bohemicum. They are best developed in its southwestern part in the Tepla-Barrandian area. In other parts of the Bohemicum they are largely covered by Permo-Carboniferous and Upper Cretaceous sediments. In the Saxothuringicum and Lugicum they separate and partly envelope the Moldanubian blocks.

In the Bohemicum area the Brioverian volcano-sedimentary sequences have been divided into three parts (Figure 6.3):

(1) The Lower Brioverian, containing abundant tholeiitic basic volcanics.

(2) The Middle Brioverian, with calc-alkaline acid and intermediate volcanics.

(3) The Uppér Brioverian, which does not contain any volcanics.

A two-fold subdivision which assigns the Middle Brioverian to either the Lower Brioverian (Chaloupsky, 1988) or the Upper Brioverian (Zoubek, 1988b) has also been used.

6.4.1 *The Lower Brioverian*

The Lower Brioverian volcano-sedimentary sequence in the Tepla-Barrandian area includes a substantial part of the spilitic and pre-spilitic group of Kettner (1917) and the Kralup-Zbraslav group of Masek and Zoubek (1980). The sequences are weakly metamorphosed and attain a thickness of c. 5000 m (Holubec, 1966). They are composed of monotonous alternating greywackes, siltstones and shales, with greywackes predominant. Black pyrite-bearing shales and silicites with stromatolites (Pacltova and Pouba, 1975) are also present. Interlayered basic volcanics and volcaniclastic rocks are concentrated in several belts (Rohlich, 1965; Holubec, 1966; Chab and Pelc, 1968; Fiala, 1978). The geochemistry of both volcanics and sediments indicates that the Lower Brioverian sequences in the Bohemicum were deposited on thinned continental and/or oceanic crust (Fiala, 1978; Jakes et al., 1979; Waldhausrova, 1984).

In other sectors of the Bohemicum the Lower Brioverian is represented by monotonous phyllites with metabasalts (now metamorphosed to greenschists or amphibolites). In the Zelezne hory Mts the Lower Brioverian sequences include deposits of carbonate Fe–Mn ores and iron sulfides (Fiala and Svoboda, 1956).

In all the Bohemicum areas the metabasalts are concentrated in the lowermost levels of the Brioverian and located along tectonic zones near the boundary with the adjacent Moldanubian units. In these areas they are more intensely metamorphosed and are amphibolitic, partly eclogitic. They are interlayered with mica schists and paragneisses, locally with acid meta-volcanics, leptynites or orthogneissic rocks and associated with gabbros, ultramafic rocks (peridotites and serpentinites) and trondhjemites.

These amphibolite-gabbro-serpentinite assemblages have been sometimes interpreted in terms of a dismembered ophiolite complex. They are well indicated by high gravity and magnetic anomalies (Bucha et al., 1989). In the Bohemicum they include amphibolite gabbro and serpentine complexes near Erbendorf in Oberpfalz (Stettner, 1988), Marianske Lazne, Pobezovice, Kdyne, Ransko and Letovice (Misar et al., 1984; Vejnar, 1984). Nove Mesto (Opletal, 1980) and in the basement to the Permo-Carboniferous sediments south of the Krkonose Mts (Chaloupsky, 1989). They also occur in the Saxothuringicum—in the Central Saxon Lineament near Munchenberg (the 'Hangendserie'; Stettner, 1988; allochthonous after Franke, 1984) and near Wildenfels and also in the Lugicum—in the fault zones rimming the Moldanubian Sowie Gory Gneiss Massif (Late Proterozoic according to Oberc, 1972; Early Palaeozoic according to Pin et al., 1988), the amphibolite complex near Klodzko (Vendian according to Wojciechowska and Gunia, 1988) and in the Stare Mesto Zone.

The Brioverian sequences in the Saxothuringicum and Lugicum outside the major tectonic zones are mainly greywackes with sporadic volcanic rocks. Only the oldest of them may be of Lower Brioverian age; these include the Gorlitz Group in Lusatia (Hirschmann, 1966; Brause, 1970) and the Machnin Group in the northern Jested Mts (Chaloupsky, 1989).

6.4.2   The Middle Brioverian

In the Barrandian area southeast of Prague, the Middle Brioverian is represented by the Davle Formation which is a c. 1000 m thick assemblage of volcanic and volcaniclastic rocks of andesite and dacite to Na-rhyolite composition. Acid to intermediate volcanic rocks of calc-alkaline trend succeed Lower Brioverian sequences with tholeiitic basic volcanics (Rohlich and Fediuk, 1964; Rohlich, 1965; Masek, 1981; Waldhausrova, 1984). They are overlain by a distinct, c. 50–200 m-thick horizon of black siliceous shales (the Lecice Member of Rohlich, 1965) and at higher levels by the Upper Brioverian flysch.

Weakly metamorphosed tuffitic shales, greywackes, andesites and dacites, overlain by black shales and silicites are representative of the Middle Brioverian in the Zelezne hory Mts (post-deposit division of Fiala and Svoboda, 1956).

In other sectors of the Bohemian Massif the position of acid to intermediate volcanics within the Lower and Middle Brioverian sequences is not so clear — near Vitanov (Vachtl, 1971) in the basement of overstep sequences in northern Bohemia (Chaloupsky, 1989), in the lower part of the Katzhute Group near Schwarzburg (Bankwitz and Bankwitz, 1975), and in the Leszczyniec Volcanic Formation east of the Krkonose Mts (Lower Palaeozoic after Narebski et al., 1986; Upper Proterozoic after Chaloupsky, 1989).

The Middle Brioverian black shales and silicites in the Barrandian and Zelezne hory areas may have their stratigraphic equivalents in the black shales mentioned by Falk (1974) and Hofmann et al. (1988) from the lower part of the North Saxon Group in the Saxothuringicum. Laminated black slates and phyllites have also been identified in the Jested and Zelezny Brod areas (the lower part of the Radcice Group; Chaloupsky, 1989) and in the Kaczawa Mts (the Radzimowice schists; Teisseyre, 1980).

All these black pelitic or siliceous rocks lie close to the base of the Upper Brioverian successions. In the Zelezny Brod area they unconformably overlie the much older Moldanubian schists.

6.4.3   The Upper Brioverian

The Upper Brioverian sequences are restricted to two areas: the Barrandian area southeast of Prague (Stechovice Group) and to a zone stretching in the Saxothuringicum-Lugicum from Schwarzburg over Leipzig, Elbe/Labe zone up to Lusatia (North Saxon Group) with an eastern continuation to the Jested and Kaczawa Mts areas.

The Stechovice Group (post-spilitic Group of Kettner, 1917) is c. 4 km thick (Masek, 1981) and composed of alternating shale, siltstone, greywacke and several horizons of intraformational conglomerate. The conglomerates contain pebbles of previously deposited Brioverian sediments and volcanics together with some granites and low-grade metamorphic schists (Fiala, 1948). The sediments are weakly metamorphosed and display well preserved internal structures characteristic of turbidite sedimentation.

The assignment of the Sovolusky Group in the Zelezne hory Mts (the 'Eocambrian' of Fiala and Svoboda, 1956) to the Upper Brioverian is questionable. The group may be of Lower Cambrian age.

In the Saxothuringicum, the Upper Brioverian flysch sequences of the North Saxon Group also commonly comprise conglomerates (near Schwarzburg, Leipzig, Weesenstein and Kamenz; Hirschmann, 1966; Lorenz and Burmann, 1972; Bankwitz and Bankwitz, 1975; Frischbutter, 1982; Hofmann et al., 1988). In the Jested and Zelezny Brod areas the Upper Brioverian is represented by phyllites, metagreywackes and meta-conglomerates in the lower part of the Radcice Group (Chaloupsky, 1989).

6.5 Unit boundaries

The oldest K–Ar ages obtained from the Lower Brioverian metavolcanics in the Saxothuringicum and Bohemicum areas range between 790 and 650 Ma (Zoubek, 1988a and references therein) and those of the Middle Brioverian volcanics in the Barrandian area between 700 and 600 Ma (Masek and Zoubek, 1980). The lower boundary of the Brioverian in the Bohemian Massif may be thus dated at about 900–800 Ma. The present microfossil finds do not allow a more precise stratigraphic assignment or subdivision (Konzalova, 1988).

Lithostratigraphically, the lower boundary of the Brioverian can be identified with the first significant accumulation of basic volcanic rocks of tholeiitic type. These rocks are usually intensely metamorphosed (amphibolite to eclogite facies) so that their unconformable relationship to the underlying Moldanubian schists cannot be clearly defined. The boundary is mostly tectonic, stitched by plutons or covered by Permo-Carboniferous and younger platform sediments.

The appearance of acid to intermediate volcanics of calc-alkaline type as well as the wide distribution of the black shale facies in the Middle Brioverian time (around 650 Ma), marks the end of a period of extensional tectonism and the beginning of a period of deformation, metamorphic activity, and emplacement of plutonic rocks. In this early Cadomian event the Brioverian basin was dissected, the central zone of the Bohemicum started to elevate, and Upper Brioverian sedimentation migrated to its peripheries and partly over-lapped onto the Moldanubian basement rocks (Figure 6.3). The result was large-scale transgression of black shale formations and migration of the overlying Upper Brioverian flysch sediments on to the Moldanubian complexes in the adjacent Saxothuringicum and Lugicum.

The late Cadomian folding which followed the deposition of the Upper Brioverian flysch is apparent in the unconformable superposition of unmetamorphosed Cambrian and Ordovician sediments on folded and weakly metamorphosed Brioverian rocks in the Barandian area (Kettner, 1917; Dudek and Fediuk, 1955). The unconformity has also been identified in Lusatia (Hirschmann, 1966) and in the Jested and Zelezny Brod areas (Chaloupsky, 1989). It is controversial near Schwarzburg (Bankwitz and

Bankwitz, 1975) and in the Erzgebirge/Kruzne hory Mts (Hofmann *et al.*, 1988).

6.6 Cadomian deformation

Precambrian and Lower Palaeozoic rocks of the Bohemian Massif display various intensities of deformation. The most intense Variscan and Caledonian deformations are apparent in tectonic zones which separate major Moldanubian and Brioverian units. Some of these zones show a complex imbricate structure. Occasionally they have been interpreted as zones of large scale overthrusting or zones of significant strike-slip displacements. The Variscan folds in the massif strike dominantly NE–SW, the Caledonian and/or Acadian folds E–W to WNW–ESE.

Cadomian structures in the low-grade Brioverian rocks of the Tepla-Barrandian area are characterized by large cylindrical folds with well developed cleavage (Holubec, 1966). Simple fold structures are also reported from the Brioverian rocks in Lusatia (Hofmann *et al.*, 1988). The Cadomian folds in both Brioverian and Moldanubian units generally follow the basic E–W arc-like structure of the massif. Locally they adjust themselves to the pre-existing tectonic zones and irregular outlines of the Moldanubian blocks.

The structure of the higher grade Moldanubian units is generally more complicated and typically of polyphase character. There is much uncertainty in the assignment of individual deformational events to the Variscan, Caledonian, Cadomian and pre-Cadomian orogenic events (Svoboda, 1966; Oberc, 1972; Fiala, 1978; Bankwitz and Bankwitz, 1982; Stein, 1988).

Pre-Cadomian (Grenville) folds have been identified in the Moldanubian unit of the Krkonose Mts. They are represented by variously deformed open folds and large anticlinorial and synclinorial structures of NNE–SSW to N–S trend (Chaloupsky, 1989). Also in the Moldanubicum and Orlicke hory-Klodzko areas folds of a roughly submeridional trend have been interpreted as pre-Cadomian (Benes, 1964; Fajst, 1976). The eastern boundary of the area of Grenville regeneration may correspond to the NNE–SSW trending, rejuvenated Moravicum-Silesicum Tectonic Zone on the western boundary of the Brunovistulicum.

6.7 Cadomian metamorphism and plutonism

Cadomian and Variscan tectono-thermal events played the dominant role in the polyphase metamorphic evolution of the Bohemian Massif. Cadomian metamorphism is apparent in the Tepla-Barrandian area where folded and weakly metamorphosed Brioverian sequences are overlain unconformably by unmetamorphosed Cambrian and younger Palaeozoic sediments. In this area the Brioverian slates are largely metamorphosed under prehnite-pumpellyite

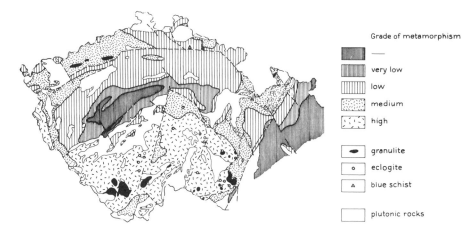

Grade of metamorphism

—

very low

low

medium

high

granulite

eclogite

blue schist

plutonic rocks

Figure 6.4 Regional metamorphism of Precambrian and Lower Palaeozoic rocks in Czechoslovakia (overstep sequences are stripped away).

facies conditions (Figure 6.4). As the surrounding Moldanubian units are approached, metamorphism increases to the greenschist facies and then rapidly up to the amphibolite facies through well defined chlorite, biotite and garnet zones (Vejnar, 1984). In the lowermost Brioverian and Moldanubian units the higher grade metamorphism is characterized by staurolite- and kyanite-bearing assemblages and locally by the development of high pressure eclogitic rocks and granulites. At these levels however, an earlier medium pressure metamorphism is largely overprinted and partly obscured by later low pressure metamorphism characterized by assemblages with cordierite and sillimanite and a regional migmatization. In view of its postkinematic character and given the abundance of Carboniferous radiometric ages in the higher grade metamorphic Moldanubian rocks, the low pressure metamorphism and migmatization is often assumed to be of Variscan age (Blumel and Schreyer, 1976; Seidel *et al.*, 1980; Van Breemen *et al.*, 1982; Wagener-Lohse and Blumel, 1986; Suk, 1988).

While the metamorphism of the Lower Brioverian rocks may reach amphibolite and locally even eclogite facies conditions, the metamorphism of the Upper Brioverian sediments is always very weak, not exceeding the chlorite zone. This might indicate that the climax of the Cadomian medium-pressure metamorphism occurred during the early Cadomian orogenic event, during which most of the Cadomian gneisses, granitoid and gabbro-diorite plutonic rocks may have been generated. Many of them appear to be autochthonous or para-autochthonous, formed *in situ* by anatectic granitization of surrounding supracrustal rocks. Their characteristics are readily related to the rock unit and type of crust in which they are found: the gneisses are concentrated in the Moldanubian Varied Group, granite-granodiorite-quartz diorite suites near the boundary of the Moldanubian and Brioverian units, gabbros and diorites in the lowermost Brioverian amphibolite complexes. The last granitoid rocks were emplaced just after the late Cadomian folding near the Proterozoic-Cambrian boundary.

The Cadomian granitoid rocks associated with both orogenic phases display a wide dispersion of radiometric (mostly K–Ar) ages between 625 and 540 Ma. They commonly occur in Lusatia (Hirschmann, 1966; Brause, 1970; Haake et al., 1973; Chaloupsky, 1989) in the Bohemicum (Svoboda et al., 1966; Klominsky and Dudek, 1978; Zoubek, 1988b, and others) and may also be included in the granitoid massifs of the Brunovistulicum (Dudek, 1980), Moldanubicum and Saxothuringicum-Lugicum.

Assessment of the scope and effects of the pre-Cadomian (Grenville and/or earlier) metamorphic processes in the Moldanubian units is still problematic due to the effects of Cadomian and Variscan tectono-thermal reworking. Grauert et al. (1974) find no evidence for high-grade metamorphism within the period 1500–700 Ma, which would encompass the period of sedimentation of the Lower Brioverian, Upper Moldanubian and possibly also Middle Moldanubian units as well as the Grenville deformations. There is also no geological evidence for higher grade pre-Cadomian (Grenville) metamorphism of the Upper Moldanubian units. Radiometric ages of various relict zircons within the Central European Variscides range between 2500 and 1500 Ma (Wendt et al., 1989 and others). Nd model age data for central Europe are concentrated in the period 1700–1400 Ma (Liew and Hoffman, 1988) and might possibly define some tectono-thermal event near the Upper/Middle or Middle/Lower Moldanubian boundaries.

Variscan metamorphism variably overprints the earlier Cadomian and/or pre-Cadomian metamorphism. Early Variscan (Ligerian-Acadian–Late Caledonian) metamorphism was synchronous with Silurian to upper Devonian deformations which were intensely developed along the main tectonic zones of the Saxothuringicum and Lugicum. In the Central Saxonian Lineament it is represented by low grade metamorphic basaltic rocks with barroisitic amphiboles (Schwab and Mathe, 1981) and in the Zelezny Brod and south Krkonose areas by sporadic occurrences of glaucophane or crossite (Fediuk, 1962; Wieser, 1978). In the surrounding higher grade Precambrian metamorphics it is marked by more or less intense retrogression. Variscan heating in Carboniferous times, indicated by abundant radiometric ages (360–290 Ma), was associated with further mobilization and emplacement of granitoid rocks and a strong thermal overprinting of the Moldanubian metamorphic rocks.

Acknowledgements

I wish to thank all my 'Bohemian' colleagues for fruitful discussions, and the editors for their critical reading of the manuscript, constructive comments and corrections of my too continental English.

References

Bankwitz, E. and Bankwitz, P. (1975) Zur Sedimentation proterozoischer und kambrischer Gesteine im Schwarzburger Antiklinorium. *Z. Geol. Wiss.* **3**, 1279–1305.

Bankwitz, P. and Bankwitz, E. (1982) Zur Entwicklung der Erzgebirgischen und der Lausitzer Antiklinalzone. *Z. Angew. Geol.* **28**, 511–524.

Behr, H. J., Engel, W., Frankie, W., Giese, P. and Weber, K. (1984) The Variscan belt in central Europe: main structures, geodynamic implications, open questions. *Tectonophysics* **109**, 15–40.

Benes, K. (1964) Analyza vnitni stavby moldanubicko-assyntske hranicni oblasti pri sv. okraji moldanubickeho jadra. *Rozpr. Cesk. Akad. Ved., Rada Mat. Prir. Ved.* **74**, 1–80.

Blumel, P. and Schreyer, W. (1976) Progressive regional low-pressure metamorphism in Moldanubian metapelites of the northern Bavarian Forest, Germany. *Krystalinikum* **12**, 7–30.

Brause, H. (1970) Ur-Europa und das gefaltete sachsische Palao-zoikum. *Ber. Dtsch. Ges. Geol. Wiss., Reihe A* **15**, 327–361.

Bucha, V., Ibrmajer, J. and Blizkovsky, M., eds. (1989) *Lithosphere of Czechoslovakia.* Praha (in press).

Chab, J. and Pelc, Z. (1968) Lithology of Upper Proterozoic in the northwest limb of the Barrandian area. *Krystalinikum* **6**, 141–167.

Chaloupsky, J. (1988) Major tectonostratigraphic units of the Bohemian Massif. In *Terranes in the Circum-Atlantic Paleozoic Orogens*, ed. Dallmeyer, R. D., *Geol. Soc. Am. Spec. Pap.* **230**.

Chaloupsky, J. (1989) Geological evolution of the Krkonose-Jizerske hory crystalline complex in the Precambrian and Early Paleozoic. In *Geologie Krkonos a Jizerskych hor*, ed. Chaloupsky, J., Oblastni Reg. Geol. CSR, Ustr. Ust. Geol., Praha.

Dudek, A. (1980) The crystalline basement block of the Outer Carpathians in Moravia: Bruno-Vistulicum. *Rozpr. Cesk. Akad. Ved., Rada Mat. Prir. Ved.* **90**, 1–85.

Dudek, A. and Fediuk, F. (1955) Zur Altersfrage der Metamorphose im barrandienischen Proterozoikum. *Geologie* **4**, 397–403.

Fajst, M. (1976) Nova diskordance v prekambriu Ceskeho masivu. *Cas. Mineral. Geol.* **21**, 257–275.

Falk, F. (1974) Jungproterozoikum. In *Geologie von Thuringen*, eds. Hoppe, W. and Seidel, G., VEB Hermann Haak, Gotha Leipzig, 119–143.

Fediuk, F. (1962) Vulkanity zeleznobrodskeho krystalinika. *Rozpr. Ustred. Ustavu Geol.* 1–116.

Fediukova, E. and Dudek, A. (1977) Trace elements of the Moldanubian eclogites. *Neues Jahrb. Mineral. Abh.* **130**, 187–207.

Fiala, F. (1948) Algicke slepence ve strednich Cechach. *Sb. Statei Geol. Ustavu* **15**m, 399–612.

Fiala, F. (1978) Proterozoic and Early Paleozoic volcanism of the Barrandian-Zelezne hory zone. *Sb. Geol. Ved, Geol.* **31**, 71–90.

Fiala, F. and Svoboda, J. (1956) Problem subkambria a subkambrickeho zaledneni v Zeleznych horach. *Sb. Ustred. Ustavu Geol., Oddil Geol.* **22**, 257–309.

Fiala, J., Matejovska, O. and Vankova, V. (1987) Moldanubian granulites. Source material and petrogenetic considerations. *Neues Jahrb. Mineral. Abh.* **157**, 133–165.

Franke, W. (1984) Variszischer Deckenbau im Raume der Munchberger Gneismasse, abgeleitet aus der Facies, Deformation und Metamorphose im ungebenden Palaozoikum. *Geotektonische Forsch.* **68**, 253 pp.

Frischbutter, A. (1982) Zur prakambrischen Entwicklung der Elbezone. *Z. Angew. Geol.* **28**, 359–366.

Grauert, B., Hanny, R. and Soptrijanova, G. (1974) Geochronology of a polymetamorphic and anatectic gneiss region. The Moldanubicum of the area Lam-Deggendorf, E. Bavaria, Germany. *Contrib. Mineral. Petrol.* **45**, 37–63.

Haake, R., Herrmann, G., Palchen, W. and Pilot, H. J. (1973) Zur Altersstellung der Granodiorite der Westlichen Lausitz und angrenzender Gebiete. *Z. Geol. Wiss.* **1**, 1664–1671.

Hirschmann, G. (1966) Assyntische und variszische Bauenheiten im Grundgebirge der Oberlausitz. *Freiberg. Forschungsh. C* **212**.

Hofmann, J., Mathe, G. and Werner, C. D. (1988) Saxothuringian zone and the Central German Crystalline Zone in the German Democratic Republic. In *Precambrian in Younger Fold Belts. European Variscides, the Carpathians and Balkans*, ed. Zoubek, V., Wiley Interscience, Chichester, 119–144.

Holubec, J. (1966) Stratigraphy of the Upper Proterozoic in the core of the Bohemian Massif. *Rozpr. Cesk. Akad. Ved* **76/4**.

Hoth, K., Lorenz, W., Hirschmann, G. and Berger, H. J. (1979) Lithostratigraphische Gliederungsmoglichkeiten regionalmetamorphen Jungproterozoikums am Beispiel des Erzgerbirges. *Z. Geol. Wiss.* **7**, 397–404.

Jakes, P., Zoubek, V., Zoubkova, J. and Franke, W. (1979) Greywackes and metagreywackes of the Tepla-Barrandian Proterozoic area. *Sb. Geol. Ved, Geol.* **32**, 83–122.

Jaros, J. and Misar, Z. (1976) Nomenclature of the tectonic and lithostratigraphic units in the Moravian Svratka Dome. *Vest. Ustred. Ustavu Geol.* **51**, 113–122.

Jencek, V. and Vajner, V. (1968) Stratigraphy and relation of the groups in the Bohemian part of the Moldanubicum. *Krystalinikum* **6**, 105–123.

Kettner, R. (1917) Versuch einer stratigraphischen Einteilung des bohmischen Algonkiums. *Geol. Rundsch.* **8**, 169–188.

Klominsky, J. and Dudek, A. (1978) The plutonic geology of the Bohemian Massif and its problems. *Sb. Geol. Ved, Geol.* **31**, 47–69.

Kodym, O. (1946) Moldanubicke zona variska v Cechach. *Sb. Ustred. Ustavu Geol.* **13**, 125.

Konzalova, M. (1988) Biostratigraphy. Bohemian Massif. In *Precambrian in Younger Fold Belts European Variscides, the Carpathians and Balkans*, ed. Zoubek, V., Wiley Interscience, Chichester, 41–47.

Liew, T. C. and Hofmann, A. W. (1988) Precambrian crustal components, plutonic associations, plate environment of the Hercynian fold belt of Central Europe: indications from a Nd and Sr isotopic study. *Contrib. Mineral. Petrol.* **98**, 129–138.

Lorenz, W. and Burmann, G. (1972) Alterskriterien fur das Prakambrium am Nordrand der Bohmischen Masse. *Geologie* **21**, 405–432.

Machart, J. (1984) Ultramafic rocks in the Bohemian part of the Moldanubicum and Central Bohemian islet zone (Bohemian Massif). *Krystalinikum* **17**, 13–32.

Masek, J. (1981) Ke geologii proterozoika jv. kridla barrandienu. *Sb. Korelace Proterozoickych a Paleozoickych Stratiformnich Lozisek 6, Ustavu Geol. Ved Prirodoved. Fak. KU, Praha*, 14–28.

Masek, J. and Zoubek, J. (1980) Navrh vymexeni a oznacovani hlavnich stratigrafickych jednotek barrandienskeho proterozoika. *Vestn. Ustred. Ustavu Geol.* **55**, 121–123.

Maska, M., Matejka, A. and Zoubek, V., eds. (1960) *Tectonic Development of Czechoslovakia*. NCAV, Praha.

Misar, Z., Jelinek, E., Soucek, J. and Tonika, J. (1984) The correlation of gabbro-peridotite complex of the Bohemian Massif. *Krystalinikum* **17**, 99–113.

Narebski, W., Dostal, J. and Dupuy, C. (1986) Geochemical characteristics of lower Paleozoic spilite-keratophyre series in the western Sudetes (Poland): petrogenetic and tectonic implications. *Neues Jahrb., Mineral. Abh.* **155**, 243–258.

Oberc, J. (1972) *Budowa Geologiczna Plaski. Tektonika II*. Sudety i oszary pryzylegle. Inst. Geol. Warszawa, 307 pp.

Opletal, M., ed. (1980) Geologie Orlickych hor. *Oblastni Reg. Geol. CSR, Ustred. Ustavu Geol. Praha.*

Pacltova, B. and Pouba, Z. (1975) K otazce puvodu proterozoickych stromatolitu v Barrandienu. *Korelace Proterozoickych a Paleozoickych Stratiformnich Lozisek III*, 25–58.

Pertold, Z. and Suk, M. (1986) Precambrian of the Bohemian Massif and its metalogeny. *Int. Conf. on the Metallogeny of the Precambrian (IGCP Project 91)*, Geol. Surv., Prague, 9–22.

Pin, C., Majerowicz, A. and Wojciechowska, I. (1988) Upper Proterozoic oceanic crust in the Polish Sudetes: Nd–Sr isotope and trace element evidence. *Lithos* **21**, 195–209.

Reitz, E., Pflug, H. D. and Franke, W. (1988) *Biostratigraphie im Kristallin*. 1. KTB-Schwerpunkt-Kolloquim, Giessen.

Rohlich, P. (1965) Geologische Probleme des mettelbohmischen Algonkiums. *Geologie* **14**, 373–403.

Rohlich, P. and Fediuk, F. (1964) Profil barrandienskym algonkiem jizne od Prahy. Geologicky pruvodce. *Knih. Ustred. Ustavu Geol.*, Praha.

Schwab, M. and Mathe, G. (1981) A geological cross-section through the Variscides in the German Democratic Republic (Eastern Erzgebirge, Central Saxonian Lineament, Saxonian Granulite complex, Hartz Mountains). *Geol. Mijnbouw* **60**, 129–135.

Seidel, E., Kreuzer, H., Schussler, U., Okrusch, M., Lenz, K. L. and Raschka, H. (1980) *K–Ar*

Geochronology of the East-Bavarian Basement. 1. KTB-Schwerpunkt-Killoquim, Giessen, p. 59.

Stein, E. (1988) Die strukturgeologische Entwicklung im Ubergangsbereich Saxothuringikum/Moldanubikum in NE-Bayern. *Geol. Bavarica* **92**, 5–131.

Strettner, G. (1988) The Moldanubian region in the Bavarian segment of the Bohemian Massif (Federal Republic of Germany). In *Precambrian in Younger Fold Belts. European Variscides, the Carpathians and Balkans*, ed. Zoubek, V., Wiley Interscience, Chichester, 252–267.

Suk, M. (1974) Lithology of Moldanubian metamorphics. *Cas. Mineral. Geol.* **19**, 373–389.

Suk, M. (1988) Metamorphic history of the European Variscan belt. In *Precambrian in Younger Fold Belts. European Variscides, the Carpathians and Balkans*, ed. Zoubek, V., Wiley Interscience, Chichester, 17–35.

Svoboda, J., ed. (1966) *Regional Geology of Czechoslovakia, I., The Bohemian Massif.* NCSAV, Praha.

Teisseyre, H. (1980) Precambrian in southwest Poland. *Geol. Sudetica* **15**, 7–40.

Vachtl, J. (1971) Acid volcanic rocks of the Vitanov Group (Zelezne hory Mts.). *Acta Univ. Carol., Geol., Hejtman* Vol. **1–2**, 167–174.

Van Breemen, O., Aftalion, M., Bowes, D. R., Dudek, A., Misar, Z., Povondra, P. and Vrana, S. (1982) Geochronological studies of the Bohemian massif, Czechoslovakia, and their significance in the evolution of Central Europe. *Trans. R. Soc. Edinburgh* **73**, 89–108.

Vejnar, Z. (1965) Bemerkungen zur lithostratigraphischen Beziehung zwischen dem mittelbohmischen Algokium und dem Moldanubikum. *Neues Jahrb. Geol. Palaeomontol. Monatsh.* **2**, 102–111.

Vejnar, Z., ed. (1984) Geologie domazlicke oblasti. *Oblastni Regionalni Geol. CSR, Ustred. Ustavu Geol.,* Praha.

Vrana, S. (1979) Polyphase shear folding and thrusting in the Moldanubicum of Southern Bohemia. *Vestn. Ustred. Ustavu Geol.* **54** (2), 75–86.

Wagner-Lohse, Chr. and Blumel, P. (1986) Prograde Niederdruckmetamorphose und altere Mitteldruckmetamorphose im nordostbayerischen Abschnitt der Grenzzone Saxothuringikum/Moldanubikum. *Jahrestag. Geol. Ver. Giessen* **76**, 84–85.

Waldhausrova, J. (1984) Proterozoic volcanics and intrusive rocks of the Jilove zone in central Bohemia. *Krystalinikum* **17**, 77–97.

Wendt, J. I., Kroner, A., Todt, W., Fiala, J., Rajlich, P., Liew, T. C. and Vanek, J. (1989) U–Pb zircon ages and Nd whole rock systematics for Moldanubian rocks of the Bohemian Massif, Czechoslovakia. In *Proceedings of the First International Conference on the Bohemian Massif,* ed. Kukal, 2.

Wieser, T. (1978) Glaucophane schists and associated rocks of Kopina Mt. (Lasocki Range, Sudeten). *Mineral. Pol.* **9**, 17–40.

Wojciechowska, I. and Gunia, T. (1988) Precambrian of the Sudetes and their foreland. In *Precambrian in Younger Fold Belts. European Variscides, the Carpathians and Balkans*, ed. Zoubek, V., Wiley Interscience, Chichester, 161–179.

Zeman, J. (1984) Granulites of the Bohemian Massif related to its deep structure and development. *Krystalinikum* **15**, 81–102.

Zoubek, V. (1988a) Variscan belt. Conclusions, correlations. In *Precambrian in Younger Fold Belts. European Variscides, the Carpathians and Balkans*, ed. Zoubek, V., Wiley Interscience, Chichester, 575–605.

Zoubek, V., ed. (1988b) *Precambrian in Younger Fold Belts. European Variscides, the Carpathians and Balkans.* Wiley Interscience, Chichester.

7 Precambrian terranes in the Iberian Variscan Foldbelt

C. QUESADA

7.1 Introduction

Precambrian rocks largely occur in the Iberian Peninsula within the so-called Iberian or Hesperian Massif, which occupies the western half of the peninsula and constitutes the southwesternmost and largest occurrence of the European Variscan Foldbelt. The Iberian Massif has been recently interpreted as consisting of several tectonostratigraphic terranes (Figure 7.1)

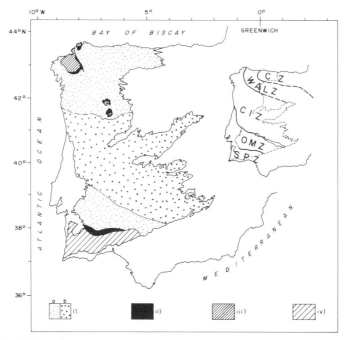

Figure 7.1 Palaeozoic tectonostratigraphic terranes in the Iberian Massif: i) Iberian Terrane (a = imbricated margins, b = autochthonous/parautochthonous core); ii) Oceanic exotic terranes; iii) Continental suspect terranes of northwest Iberia; iv) South Portuguese Terrane. **Inset map**: Zonal subdivision according to Julivert *et al.* (1974): CZ = Cantabrian Zone, WALZ = West Asturian-Leonese Zone, CIZ = Central Iberian Zone, OMZ = Ossa Morena Zone, SPZ = South Portuguese Zone.

that were accreted together during the Variscan orogeny in late Palaeozoic times (Ribeiro *et al.*, 1987). These include:

(1) The so-called Iberian miogeocline or Iberian terrane; a continental block which occupies an area greater than 80% of the entire Massif and shows an autochthonous/parautochthonous core rimmed by intensely imbricated, wide marginal areas. The Precambrian basement of this continental terrane, which is considered as the key reference unit for terrane analysis at peninsular scale, consists of several sequences collectively showing evidence of having undergone a late Proterozoic Pan-African or Cadomian history. This links the evolution of the Iberian terrane to that of Gondwanaland at least in late Proterozoic times.

(2) An ophiolitic terrane or terranes, represented by Palaeozoic ocean floor and overlying sedimentary sequences, that have been provisionally grouped together, and are interpreted as representing dismembered pieces of an ophiolitic thrust sheet obducted eastward onto the Iberian terrane during the Variscan convergence (Ribeiro *et al.*, 1989). These ophiolites are regarded as remnants of a Palaeozoic ocean lying west of the Iberian terrane, and formed as a result of a rifting process that is very well documented in that terrane during the Cambrian–Silurian timespan (Iglesias *et al.*, 1983; Quesada *et al.*, 1987).

(3) Unnamed continental units of unknown palinspastic origin which constitute thrust sheets sitting on top of ophiolitic nappes in northwest Iberia (Cabo Ortegal, Ordenes, Bragança and Morais massifs). They are mostly composed of Precambrian basement rocks locally overlain by lower Palaeozoic clastic cover sequences. The limited areal extent of these sequences together with the scarcity of geochronological data prevent their correlation with basement sequences in the Iberian terrane or elsewhere. No late Proterozoic Pan-African ages have ever been reported for these rocks. Nevertheless, *c.* 2.5–2 Ga old zircons have been found in them (Kuijper, 1979; Kuijper *et al.*, 1982); this may be interpreted as suggesting an African linkage in late Archaean–early Proterozoic times. Whether these units represent a block that was rifted apart from the Iberian continental terrane during the opening of the ocean where the ophiolites described above originated still remains a matter for speculation.

(4) The so-called South Portuguese terrane which occurs in Southern Iberia. This is another continental block separated from the Iberian terrane by an intervening oceanic unit: the Pulo do Lobo unit which includes the Beja-Acebuches ophiolite (Munha *et al.*, 1986). The rock sequences recorded in the South Portuguese terrane correspond exclusively to late Palaeozoic cover successions.

The aim of this chapter is to provide an overview of the various Precambrian sequences which occur in the continental terranes referred to above, with the obvious exception of the South Portuguese terrane, in order to put some constraints on their respective evolutionary histories prior to their accretion in Palaeozoic times. This is not a straightforward task mainly due to the dispersion and generally severe tectono-thermal overprint that the Variscan orogeny has imposed on the original relationships, together with the scarcity of reliable geochronological and geochemical data which are crucial for the understanding of highly complex rocks such as these.

7.2   Precambrian sequences and history of northwest Iberia suspect terranes

The presence of polymetamorphic rocks attributed to the Precambrian has been known in northwest Iberia since early in the 19th century (Schulz, 1835; Macpherson, 1881), and their distribution and petrography are very well known after the works by Parga Pondal (1956, 1960, 1963). However, it was not until a fundamental paper by Ries and Shackleton (1971) that their interpretation as dismembered pieces of an allochthonous thrust plate became widely accepted. They are now currently regarded as defining the uppermost allochthonous sheets which overlie an ophiolite nappe in the large, structurally complex klippes of the Cabo Ortegal, Ordenes, Bragança and Morais massifs (Figure 7.2). These klippes of continental and oceanic suspect terranes overlie extremely imbricated Palaeozoic cover sequences belonging to the Iberian terrane (Figures 7.2 and 7.3).

The best known and simplest structural situation occurs in the Morais

Figure 7.2   Location of Palaeozoic suspect/exotic terranes in the northwest Iberian Massif (1 = Ocean-derived units, 2 = Continental units). Adapted after Iglesias *et al.* (1983).

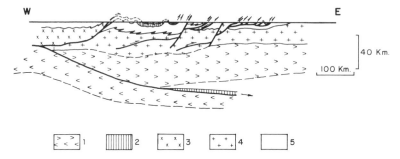

Figure 7.3   Schematic structural section across the northwest Iberian Massif showing the disposition of allochthonous exotic terranes onto imbricated Iberian Terrane miogeocline sequences (after Iglesias *et al.*, 1983). 1 = Upper mantle; 2 = Oceanic crust; 3 = Basement of continental exotic terrane; 4 = Basement of the Iberian terrane; 5 = Palaeozoic cover sequences.

Massif in Northern Portugal, with which a correlation of the rock sequences of the other massifs can be established in spite of their more complex internal arrangement. The upper allochthonous nappe complex in this massif shows what seems to be a telescoped section across a continental lithospheric block (Ribeiro *et al.*, 1987); the major tectonostratigraphic elements of which are shown schematically on Figure 7.4 along with their interpretation according to Ribeiro (1983) and Ribeiro *et al.* (1987).

In the absence of any radiometric dating the most important features to be taken into account in order to distinguish Precambrian from Palaeozoic

AGE	SEQUENCE	INTERPRETATION
LOWER PALEOZOIC	Lagoa Micaschist	COVER SEQUENCE
PRECAMBRIAN	Lagoa Augengneiss	UPPER CRUST
PRECAMBRIAN	Mafic Granulites	LOWER CRUST
PRECAMBRIAN	Garnet bearing Lherzolites	UPPER MANTLE
LOWER PALEOZOIC	Morais Ophiolite	OCEANIC CRUST

Figure 7.4   Tectonostratigraphic cartoon showing the internal structure and interpretation of the upper allochthonous unit (continental suspect terrane) in the Morais Massif (based on Ribeiro, 1983, and Ribeiro *et al.*, 1987).

tectono-sedimentary/tectono-thermal processes (Ribeiro *et al.*, 1987) are: (1) the existence of a foliation in the Lagoa augen-gneisses (Figure 7.4) that predates emplacement of mafic dykes and sills that also intrude the overlying lower Palaeozoic micaschists (Mibofanov and Timofeev, in Chacón, 1979). These mafic dykes were subsequently affected by Variscan deformation and transformed into amphibolites; and (2) the existence of at least two phases of penetrative deformation of the mafic granulites and metaperidotites (Figure 7.4) developed under granulite facies metamorphic conditions, prior to the formation of tensional veins in which hornblende crystals grew with their long axes parallel to the stretching lineation of the Variscan structures.

The first consequence of these observations is the confirmation of the basement nature of the sequence except for the lower Palaeozoic Lagoa micaschists which represent a sedimentary cover. Other straightforward conclusions are: (1) the existence of polyphase deformational events of Precambrian age; (2) the existence of granulite facies Precambrian metamorphism in these sequences; and (3) the existence of Precambrian granitic plutonism in this sequence. However, no age constraints can be put to these processes, preventing therefore, their correlation with similar processes identified in the basement of the Iberian terrane. The reference to *c.* 2.5–2 Ga-old zircons found in rocks of the Ordenes and Cabo Ortegal massifs (Kuijper, 1979) suggests that a piece of late Archaean/early Proterozoic continental crust is present or has contributed to the formation of those rocks. Similarly aged zircons have been also reported from the Iberian terrane (Schäfer *et al.*, 1988, 1989). This coincidence can be perhaps considered as a basis for the correlation of the basement in both terranes, but this is clearly a point requiring further research.

7.3 Stratigraphy of Precambrian sequences in the Iberian terrane

Precambrian rocks are widely exposed throughout the Iberian terrane (Figure 7.5) in the cores of broad Variscan antiformal structures. They have been severely overprinted by late Palaeozoic Variscan tectono-thermal processes which, together with the general scarcity in Iberia of reliable geochemical and geochronological data, add a great deal of difficulty to the unravelling of their history prior to the Palaeozoic and, locally, even of their detailed stratigraphy. In general terms, all the Precambrian successions in the Iberian terrane show evidence of having been more or less affected by late Proterozoic orogenic processes, roughly time equivalent to the Pan-African or Cadomian orogenies. The relationships of the various Precambrian sequences to the Cadomian deformational events have served as a basis for their classification (Quesada *et al.*, 1987; Quesada, 1990) into: (1) pre-orogenic, and (2) syn-orogenic. With regard to their areal distribution, the pre-orogenic sequences are restricted to the southern and westernmost parts

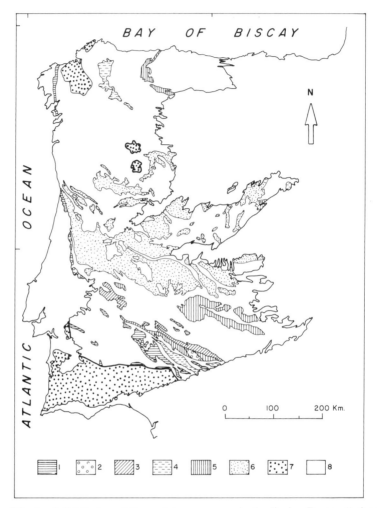

Figure 7.5   Areal distribution of Precambrian sequences in the Iberian Terrane (1, 2 and 3, pre-orogenic sequences: 1 = Valencia de las Torres-Cerro Muriano Supergroup; 2 = Sierra Albarrana Group; 3 = complex imbricated belt in which both sequences occur along with intervening serpentinite sequences. 4, 5 and 6, syn-orogenic sequences: 4 = volcano-sedimentary; 5 = flysch; 6 = undifferentiated late-Proterozoic flysch and Lower Cambrian rift sequences; 7 = suspect terranes other than the Iberian terrane; 8 = post-Palaeozoic sequences).

of the terrane whereas the syn-orogenic successions are widespread throughout (Figure 7.5).

7.3.1   Pre-orogenic sequences

These are best exposed and less penetratively reworked by the Variscan orogeny in the southern part of the terrane: the Ossa-Morena zone (Lotze,

1945; see inset map on Figure 7.1). Three pre-orogenic sequences are recognized in this zone, separated from each other by tectonic contacts. They usually appear structurally superposed with the following arrangement (from bottom to top):

(1) The so-called Sierra Albarrana Group (Apalategui et al., 1985). This succession consists of a metasedimentary shelf sequence, the base of which is never exposed. Although some discrepancy about its detailed stratigraphy exists (Apalategui and Higueras, 1983), a widely accepted interpretation (Garrote, 1976; Quesada et al., 1987; Garrote et al., 1989) is that of a lower metapelitic succession, overlain by a c. 300 m-thick quartzitic (mostly feldspathic) formation, in which pipe-like burrows have been locally found. This is in turn succeeded by a thick (> 1000 m) Al_2O_3-rich micaschist formation with minor intercalations of feldspathic sandstones, amphibolites and calcsilicate layers. Sedimentary features in the quartzite formation (trough and planar cross-bedding, current ripples and, locally, intense burrowing) along with a sporadic presence of clast-supported conglomerates suggest a shallow marine, shelf sedimentary environment for the deposition of this sequence.

No fossils have ever been found in it. Its probable Precambrian age (Delgado, 1971) has been discussed by Apalategui and Higueras (1983) on the basis of its (ambiguous) relationship with Cambrian rocks at one locality. Nevertheless, the recent publication of c. 600 Ma U–Pb radiometric results from zircons recovered in one of the granitic plutons which intrude the Sierra Albarrana Group of c. 600 Ma (Schäfer et al., 1989), confirms the Precambrian age of this series, although further precision is not yet possible.

(2) Serpentinite successions. These occur in scattered small outcrops in the northern part of the Ossa-Morena Zone, structurally interleaved between underlying Sierra Albarrana Group rocks and the overlying Valencia de las Torres-Cerro Muriano Supergroup (see below). Similar meta-ultramafic rocks occur as olistholiths and clasts within late Proterozoic syn-orogenic successions further south in the Ossa-Morena Zone (Arriola et al., 1983). The composition is mostly serpentinite, although a certain degree of zonation with chlorite, talc-carbonate and tremolite rocks exists near the margins of some outcrops that according to the alteration textures may have been derived from different ultramafic rock types (Aguayo, 1985). An Alpine-like character has been suggested for these sequences, in which lense- and pod-like chromite deposits have been locally found (Arriola et al., 1984). Recently (Eguiluz, 1987; Quesada et al., 1987) their interpretation as dismembered remnants of an obducted late Precambrian ophiolite has been proposed.

The age of these serpentinites is very poorly constrained. Indirect pieces of evidence suggest that the serpentinite unit has a minimum age of Upper Riphean, as the aforementioned olistholiths are included within overlying syn-orogenic successions (Apalategui and Quesada, 1987) that have been assigned to this age on the basis of scarce fauna (Liñán and Schmitt, 1980).

Figures 7.6 and 7.7   Variably mylonitic porphyritic orthogneisses within the Azuaga Gneisses Group.

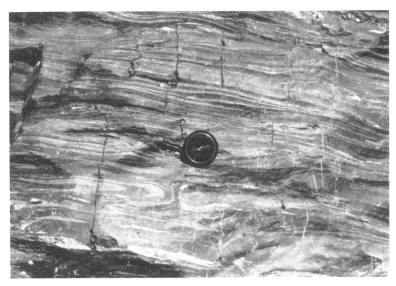

Figure 7.8   Isoclinically folded alternating plagioclase-rich and amphibole-rich layers in amphibolites of the Azuaga Gneisses Group.

(3) Valencia de las Torres-Cerro Muriano Supergroup (Quesada *et al.*, 1987). This is the most exposed pre-orogenic succession in the Iberian terrane and, where in contact, it structurally overlies the other two. It is a very complex unit with two different sequences whose mutual relationships have not yet been clearly established. The lower part (Azuaga Gneisses Group) consisting of usually high grade paragneisses, orthogneisses (Figures 7.6 and 7.7) and amphibolites (Figure 7.8), underlies (conformably or unconformably?) the Serie Negra Group, which comprises variably metamorphosed distal miogeoclinal and/or eugeoclinal rocks with abundant amphibolites (Arriola *et al.*, 1983; Eguiluz, 1987). The Azuaga Gneisses Group has been interpreted either as the reworked basement to the Serie Negra Group (Pérez, 1979; Pascual, 1981) or as a metamorphosed bimodal volcanosedimentary suite related to the early rifting of the basin within which the Serie Negra Group was subsequently laid down (Apalategui *et al.*, 1983, 1986). The latter interpretation is preferred here on the basis of field evidence, such as the lack of any metamorphic contrast between both groups where they appear in continuity, the available geochemical data which support a transitional evolution (J. Munha, pers. commun., 1988) from continental tholeiites in the Azuaga Gneisses Group (Garcia Casquero, 1989) to abyssal tholeiites in the overlying Serie Negra Group (Eguiluz, 1987), and the presence of common distinctive marker lithologies in both groups (e.g. banded graphite-rich black cherts). However, the possibility that some 'old-basement' slivers could be structurally interleaved in the succession cannot be disregarded.

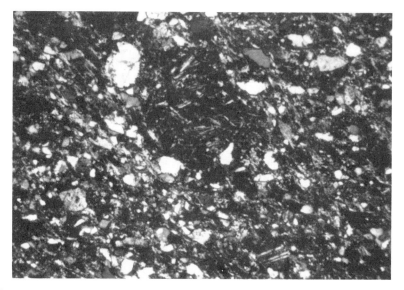

Figure 7.9   Photomicrograph of a volcaniclastic greywacke of the Serie Negra upper sequence showing clasts of andesitic rocks (the microporphyritic clast near the centre is *c.* 1.5 mm in diameter).

The Serie Negra Group consists of a thick (in excess of 5 km) succession of predominantly graphite-rich metasedimentary lithologies (pelites and grey-wackes) which define a thickening- and coarsening-upwards megasequence, in which minor lithologies are represented by amphibolites (abyssal tholeiites; Eguiluz, 1987) concentrated in the lower part of the succession, laminated black cherts and marbles. The terrigenous parts exhibit, where visible, sedimentary characteristics of turbidites and show a vertical evolution in terms of greywacke clast composition which allows separation of two parts with probable different source areas. The lower part, consisting of thinly bedded quartz-rich greywackes and pelites, was most likely derived from a mature continental source, whilst the upper part (3 km thick) consists of progressively thicker beds of massive graded greywackes including pro-gressively more abundant clasts of calc-alkaline volcanic rocks (Figure 7.9), most likely coming from an adjacent volcanic arc. This upper part should be considered as syn-orogenic *sensu stricto* and the above mentioned evolution, according to Quesada (1990), probably represents the passage from a passive margin stage with development of a miogeoclinal wedge (lower part of the Serie Negra), to a back-arc of foreland basin stage (i.e. a syn-orogenic stage). It is included here however as, according to the field evidence, both parts appear in sedimentary continuity and show the same deformational history, being separated from what is here labelled as syn-orogenic by a major tectono-thermal episode (first Cadomian event).

Again, the age of these sequences is very poorly constrained, the only fossils found in the Valencia de las Torres-Cerro Muriano Supergroup come from scattered outcrops of the Serie Negra Group that have yielded poor assemblages of achritarchs of Middle and Upper Riphéan age (Mitrofanov and Timofeev in Chacón, 1979; Chacón *et al.*, 1984).

7.3.2   Syn-orogenic sequences

Following a first Cadomian orogenic event, which is only recorded with certainty in the Valencia de las Torres-Cerro Muriano Supergroup, syn-orogenic sequences started to form. These include two main types: (1) calc-alkaline igneous suites and (2) partly coeval flysch successions. The former are restricted to the southern part (Ossa-Morena Zone, see inset map on Figure 7.1) of the Iberian terrane (Figure 7.5) and are only seen to unconformably overlie the Valencia de las Torres-Cerro Muriano Supergroup. The latter are widespread throughout the rest of the terrane, including the Ossa-Morena Zone in which they are seen to overlie every pre-orogenic sequence. The existence in northern areas within this zone of proximal turbidites (debris flows and massive turbidites) interlayered with calc-alkaline volcanics provides a spatial and temporal link between both types of syn-orogenic sequences.

(1) The calc-alkaline igneous suites consist of lava (Figure 7.10), pyroclastic (Figure 7.11), and epiclastic rocks, ranging in composition from basalt to rhyolite but with a predominance of andesites and dacites suggesting the

Figure 7.10   Pillowed andesites in the syn-orogenic volcano-sedimentary sequences.

Figure 7.11   Pyroclastic?/epiclastic breccia containing angular clasts of andesite lavas in a volcaniclastic (sand-size) andesitic matrix.

development of a mature arc on continental crust. These volcanic successions (several km thick locally) also include sedimentary intercalations of pelites and carbonates that have yielded locally very scarce fossil assemblages (algae, stromatoliths, Problematica and trace fossils) of upper Riphéan age (Liñán and Schmitt, 1980) as well as poor architarch associations of Vendian age (Palacios in Fernández *et al.*, 1983). An indirect upper limit to the age of these successions is provided by unconformable upper Vendian red beds of the Torreárboles Formation (Liñán, 1978).

Small tonalite to granite plutons intruding the pre-orogenic (and also locally the syn-orogenic) successions are widespread in the region. Their geochemistry is very similar to that of the volcanic rocks, and characteristic of subduction-related calc-alkaline arc suites (Sánchez Carretero *et al.*, 1989). These authors have also reported a progressive increasing in the K_2O content of the intermediate rocks towards the south, which could be indicative of the direction of dip of the subduction zone.

(2) The flysch-like successions, on the other hand, spread across the rest of the Iberian terrane north of the Ossa-Morena Zone (Figure 7.5). They include several depositional sequences, separated by more or less important unconformities, and reach a total thickness in excess of 5 km in central areas (Central Iberian Zone, inset map on Figure 7.1). Clastic lithologies predominate although calc-alkaline volcanic rocks and carbonates are locally important. In the Montes de Toledo area where they have been recently studied in detail (Pieren *et al.*, 1987; Vilas *et al.*, 1987; Gabaldón *et al.*, 1989) these syn-orogenic successions consist of three superposed depositional

Figure 7.12   Angular unconformity between two clastic successions belonging to the flysch syn-orogenic sequences.

sequences separated by regional unconformities (Figure 7.12). Each individual sequence shows a shallowing-upwards arrangement, and a schematic representative section includes from bottom to top: (a) basal condensed facies (pelites and cherts), known from the lowermost sequence; (b)

Figure 7.13   Metre-scale detail of the Fuentes megaturbidite; flysch syn-orogenic sequences (photo courtesy of F. Moreno).

turbidites; (c) shelf clastic and carbonate deposits, important in the inter-mediate sequence; and (d) coarse clastics of deltaic and fluviatile affinities. Several megaturbidite beds appear in the succession. The 150 m-thick Fuentes megabed (Moreno, 1977) at the base of the uppermost sequence is particularly impressive (Figure 7.13). According to the evolution of thickness and the proximal/distal character of the individual deposits, a northward migration of the whole depositional system can be envisaged. This is also supported by the sparse available palaeontologic data, which suggest the presence of upper Riphéan assemblages in the southern areas (Mitrofanov and Timofeev in Chacón, 1979) whereas a Vendian age is attributed to the fauna found in the Central Iberian zone (Palacios, 1986).

7.4   Late Precambrian tectono-thermal evolution of the Iberian Terrane: the Cadomian orogeny

Both the pre-orogenic and syn-orogenic upper Proterozoic sequences of the Iberian terrane were affected to variable degrees by orogenic processes prior to the deposition of lower Cambrian rocks which characterize the initiation of the Variscan cycle in that terrane. Before describing such processes it is perhaps worthwhile to emphasize how difficult the recognition of their results in strongly reactivated rocks like these can become, and this is especially true when dealing with structural elements or metamorphic assemblages, the interpretation of which is in most cases subjected to a certain degree of ambiguity. Nonetheless, by means of indirect evidence such as the existence of fabric bearing clasts in conglomerates (Figure 7.14) or the intrusion of lower Palaeozoic plutons (postdating the fabric in the Precambrian rocks) and, locally, by direct evidence (unconformable relationships, radiometric dating) it is now possible to recognize some relevant features of the orogenic tectono-thermal evolution of the Iberian terrane in late Precambrian times.

7.4.1 Cadomian deformational events

Two major Cadomian deformational events have been recognized in the Iberian terrane. The oldest one, as stated above, is only seen to affect with certainty the pre-orogenic Valencia de las Torres-Cerro Muriano Super-group. The second deformation affects all Precambrian sequences.

The geometrical characteristics of both deformation phases are very difficult to analyse due to the generally penetrative nature of the Variscan overprinting structures. Only in a very few, less penetratively overprinted, areas is it possible to solve this problem. This is the case for the so-called Arroyomolinos unit in the central Ossa-Morena Zone where the existence of dated intrusives (Pallares pluton: 572 ± 70 Ma; Rb–Sr whole-rock errorchron, Cueto et al., 1983; Bancarrota pluton: 505 ± 10 Ma; K–Ar hornblende age,

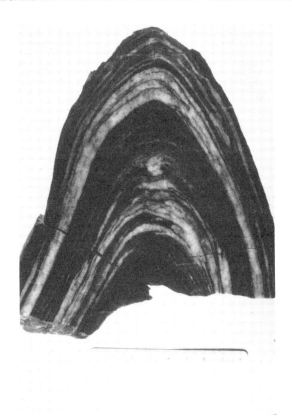

Figure 7.14   Black chert (Serie Negra Group) clast contained in basal Cambrian conglomerates, showing superposition of folding structures.

Galindo *et al.*, 1987) which developed metamorphic contact aureoles post-dating all penetrative fabrics in that unit, allows identification of all the deformation as Precambrian. In that unit the first Cadomian event, which only affects the pre-orogenic Serie Negra Group rocks, is responsible for the formation of kilometre-scale isoclinal recumbent folds (Figure 7.15) and an associated axial plane cleavage.

The geometry and characteristics of the second Cadomian deformation are largely variable from one place to another. It developed wide southwest-verging asymmetric folds in the Arroyomolinos unit whose interference pattern with the F1 structures can be seen on Figure 7.15. It formed a very penetrative flat-lying schistosity associated with isoclinal recumbent folds (only visible at meso- and microscopic scales) in the Sierra Albarrana Group rocks in the northern part of the Ossa-Morena Zone. The imbrication of pre-orogenic sequences in that area (see above) has been tentatively correlated with this second Cadomian deformation phase (Quesada *et al.*, 1987). Finally,

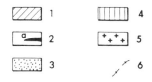

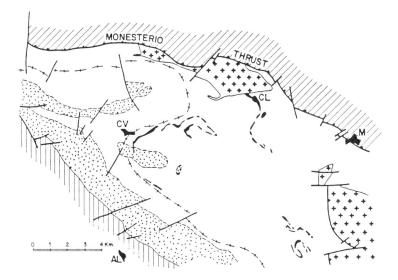

Figure 7.15   Sketch map of the central part of the Arroyomolinos unit (Monesterio Antiform) showing unconformable relationships between syn-orogenic volcanosedimentary sequences and the pre-orogenic Serie Negra Group as well as geometrical characteristics of the F1 fold structures as shown by black chert marker horizons in the latter (1 Monesterio Unit; 2, 3 and 4 Arroyomolinos Unit; 2 Serie Negra Group, a = black cherts; 3 syn-orogenic volcanosedimentary sequences; 4 Lower Cambrian sequences; 5 intrusive rocks; 6 biotite isograd).

this second deformation is responsible for the single folding phase affecting the syn-orogenic sequences north of the Ossa-Morena Zone, which evolves from northeast-verging to upright in a northeastward direction; the associated cleavage (most commonly pressure-solution cleavage) is restricted to the southernmost part of the Iberian terrane to the north of the Ossa-Morena Zone.

7.4.2   Cadomian metamorphism

Again the only areas within the Iberian terrane in which metamorphic processes associated with the Cadomian orogeny can be undoubtedly distinguished from the Variscan ones, correspond to its southernmost unit: the Ossa-Morena Zone. It is worthwhile to mention that in the authochthonous

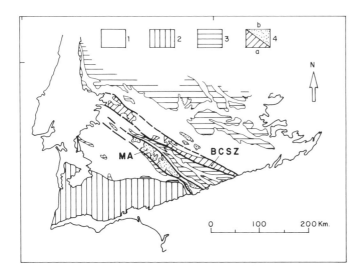

Figure 7.16 Spatial distribution of Cadomian metamorphism in the southern Iberian terrane: 1 Palaeozoic rocks in the Iberian terrane and recent cover of the Iberian Massif; 2 Palaeozoic suspect terranes other than the Iberian terrane; 3 low and very low grade Cadomian metamorphism; 4 medium and high grade Cadomian metamorphism: a = HP/HT belt, b = LP/HT belt, BCSZ = Badajoz-Córdoba shear zone, MA = Monesterio antiform.

core of the Iberian terrane (see Figures 7.1 and 7.5) which presents optimal conditions for the recognition of Precambrian processes in the basement, only anchymetamorphic to lower greenschist grade, syn-orogenic Precambrian flysch successions occur.

In the Ossa-Morena Zone most of the Precambrian occurrences appear under low and very low grade metamorphic conditions. Two belts exist, however, in which higher metamorphic regimes (amphibolite and granulite grades) were attained during Cadomian orogenic processes, as demonstrated by similar arguments discussed in the previous paragraph. In the northern belt (Figure 7.16), which roughly coincides with the well known Badajoz-Córdoba shear zone, a very complex Precambrian metamorphic evolution has been recently described by Mata and Munha (1986). In summary, they demonstrated that the P-T conditions in this zone evolved from an initial high-pressure stage resulting in eclogite formation, followed by a period of thermal reequilibration in which granulites were formed, and a subsequent interval of essentially isothermal decompression in which migmatization took place in many places, followed finally by a general retrogression, most likely related to tectonic uplift and erosion. A recently published age of 617 ± 6 Ma (U–Pb zircon in granulite; Schäfer et al., 1989) may be a good estimate for the age of the granulite facies part of the P-T path.

All the metamorphic evolution described above which can be interpreted in terms of tectonic processes resulting from a continental collision event

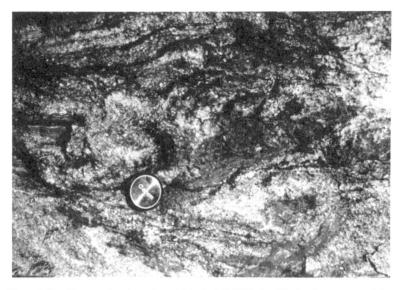

Figure 7.17   Monesterio migmatites within the LP/HT belt of Cadomian metamorphism.

followed by a period of crustal thickening, in turn succeeded by a period of upthrusting (Quesada and Munha, 1989), is only seen in rocks of the pre-orogenic Valencia de las Torres-Cerro Muriano Supergroup. The structurally underlying Sierra Albarrana Group shows a much simpler picture: a single Barrovian-type metamorphic event reaching locally lower amphibolite grade, but more commonly greenschist facies which could probably have been developed congruently with the final upthrusting of the overlying unit.

The other belt with medium and high grade metamorphic rocks (Monesterio Antiform, Figure 7.16) also shows a relatively simple picture. There the pre-orogenic Serie Negra Group displays several thermal dome metamorphic structures with a progression from the chlorite zone, through the biotite, andalusite, cordierite and sillimanite zone, to the sillimanite/ K-feldspar zone with general migmatization (Figure 7.17). The core of one of these domes is actually comprised of a cordierite-andalusite-sillimanite-bearing anatectic granodiorite (Monesterio Granodiorite; Quesada, 1975) that has been recently dated by the Rb–Sr method at 550 ± 16 Ma (Quesada *et al.*, unpublished results).

7.4.3   Cadomian igneous activity

Apart from the bimodal volcanic intercalations within the pre-orogenic Valencia de las Torres-Cerro Muriano Supergroup with continental to oceanic rifting affinities, and the thick syn-orogenic volcano-sedimentary piles described above which show characteristics typical of orogenic volcanic arc suites on continental crust, other igneous rocks correspond to two distinct types of plutonic rocks. The most widely represented type consists of calc-alkaline diorites, tonalites, granodiorites and granites, genetically related to the syn-orogenic volcanic rocks which appear in numerous generally small scattered plutons, intruding both pre- and syn-orogenic successions. Common features of all of these plutons are the presence of a generalized propylitic hydrothermal alteration, and the abundance of copper ore prospects. They are frequently deformed showing a brittle to semibrittle fabric which relates to the second Cadomian event as demonstrated by unconformable relationships, in many cases, with uncleaved Lower Cambrian sedimentary rocks. The occurrence of these plutonic rocks is restricted to the areas of volcano-sedimentary syn-orogenic successions, i.e. the Ossa-Morena Zone and adjacent areas in the southern part of the Iberian terrane.

The second type of Precambrian plutonic rocks is restricted to the areas of high grade metamorphic rocks (Figure 7.16) and consists of, volumetrically minor, anatectic per-aluminous two-mica granites, aplites and rare Al-silicate bearing granodiorites such as the Monesterio Granodiorite referred to above.

Some 'Pan-African' ages have been published for granitic rocks *sensu lato* in northern areas of the Iberian terrane (Miranda do Douro orthogneiss: 618 ± 9 Ma, U–Pb zircon, Lancelot *et al.*, 1985; Foz do Douro orthogneiss: 604 Ma, Rb–Sr, Andrade *et al.*, 1983; Eastern Sistema Central Augen-gneiss plutons: *c.* 560 Ma, U–Pb zircon, Bischoff *et al.*, 1986) but they are generally so severely overprinted by Variscan tectono-thermal events that their original relationships with their respective country rocks are very ambiguous.

7.4.4   The Cadomian orogeny in the Iberian terrane from a terrane perspective

Any attempt to analyse the various Precambrian sequences which comprise the Palaeozoic Iberian terrane within the perspective of the tectonostratigraphic terrane theory (Coney *et al.*, 1980), must face the numerous problems and limitations imposed by the severe tectono-thermal reactivation that these rocks suffered as a consequence of subsequent tectonic processes in Palaeozoic times. This resulted in many cases in dispersion of the original relationships, penetrative deformation, and resetting of the metamorphic and isotopic systems, among others. Nevertheless, the new data made available in the last decade, as shown in previous paragraphs, allows a preliminary approach to be made, bearing in mind that the results of such an analysis must be tentative.

The first analysis of Precambrian sequences in the Iberian terrane was made by Quesada *et al.* (1987). The main point of their argument lies in the existence in the northern Ossa-Morena Zone of a structural superposition of three Precambrian sequences with very different stratigraphy (the inter-

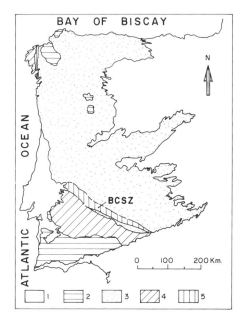

Figure 7.18   Spatial distribution of Precambrian terranes within the Palaeozoic Iberian terrane (1 post-Palaeozoic cover; 2 undifferentiated Palaeozoic suspect terranes. 3, 4 and 5 Precambrian terranes: 3 Sierra Albarrana terrane; 4 Valencia de las Torres-Cerro Muriano Arc terrane; 5 cryptic suture zone with imbrication of both 3 and 4 and intervening Precambrian ophiolites, BCSZ = Badajoz-Córdoba shear zone).

mediate one being probably ophiolitic in nature) and very different tectono-thermal history (the uppermost one showing evidence of a high pressure metamorphic event). The serpentine unit was interpreted as a remnant of a dismembered ophiolite, therefore representing a cryptic suture. A straight-forward conclusion of this interpretation is the consideration of the two units separated by the ophiolite as belonging to two different continental blocks, i.e., two different tectonostratigraphic terranes. The serpentinite sheet itself is an exotic terrane with respect to the other two. Quesada *et al.* (1987) labelled such terranes by the name of the corresponding tectono-sedimentary miogeo-clinal sequences, i.e. Valencia de las Torres-Cerro Muriano terrane and Sierra Albarrana terrane, leaving the oceanic terrane unnamed. The Cadomian orogeny is held responsible for the amalgamation of these terranes in a complex process involving subduction and subsequent continent-continent collision.

The likely areal distribution of the three Precambrian terranes which constitute the basement of the Palaeozoic Iberian terrane is shown on Figure 7.18 although a great deal of uncertainty exists in northwestern areas due to the strongly penetrative nature of the Variscan overprint and the abundance of late Palaeozoic intrusives there. A summary of the relevant features of the three terranes is presented below together with the constraints that they impose on the nature of the amalgamation process.

The continental nature of both the Valencia de las Torres-Cerro Muriano and Sierra Albarrana terranes is indicated by their respective sedimentary and/or igneous pre-orogenic records, despite the fact that no direct evidence of old basement has ever been found in either of them, and the general lack of accuracy of the current chronostratigraphic knowledge. A common signature is, however, shown by the U–Pb and Sm–Nd systematics (Nägler *et al.*, 1988, 1989; Schäfer *et al.*, 1988, 1989) which indicates the contribution of a late Archaean/early Proterozoic continental crust to the upper Precambrian and Palaeozoic evolution of both terranes.

The pre-orogenic upper Proterozoic history of both continental terranes includes a period of miogeocline growth (sea-floor spreading?) postdating what has been interpreted as a rifting sequence, only recorded in the Valencia de las Torres-Cerro Muriano terrane (Azuaga Gneisses Group). Whether this rifting event was responsible for the opening of the ocean where the serpentines were formed is as yet a matter for speculation. The same is true with regard to the consideration of the Sierra Albarrana terrane as a hypothetical counterpart, across that ocean, of the Valencia de las Torres-Cerro Muriano terrane. In the absence of any opposing evidence, and bearing in mind the possibility that large-scale strike-slip displacements could well have taken place, such interpretations are here accepted as a working hypothesis.

Several problems arise when trying to interpret what has been called the first Cadomian deformational event, which is only recognized in the Valencia de las Torres-Cerro Muriano terrane, as well as the latest part of the terrane pre-orogenic sedimentary evolution (Serie Negra Group upper sequence). Both features have been tentatively interpreted as representing a period of magmatic arc emergence and subsequent collision with a continental margin, presumably the Valencia de las Torres-Cerro Muriano terrane margin (Quesada, 1990). Unfortunately, no traces of such an arc have as yet been found, except for indirect evidence such as calc-alkaline igneous clasts within upper Serie Negra sequence conglomerates and greywackes. The possibility that this first Cadomian deformation could represent an early phase of the continent-continent collision that is neatly marked by the second Cadomian deformation is hardly likely since, postdating the first deformation, an extensive phase of magmatic arc development as shown by the syn-orogenic volcano-sedimentary sequences and related plutonic rocks, clearly suggests a period of active subduction beneath the Valencia de las Torres-Cerro Muriano terrane.

The Cadomian orogenic evolution of the Palaeozoic Iberian terrane, as it has been previously advanced, involved an initial phase of subduction and a subsequent phase of continent-continent collision after consumption of the intervening ocean. The subduction phase was characterized by the building up of an arc onto the overriding plate (Valencia de las Torres-Cerro Muriano terrane). An Andean-type active margin is regarded as the most likely environment during this phase. A southward dip of the subduction zone is suggested by the areal distribution of syn-orogenic sequences as well as the variation in K_2O content across the zone.

The second Cadomian phase most likely represented the collision between both continental blocks and resulted in the underthrusting of the Sierra Albarrana terrane beneath the Valencia de las Torres-Cerro Muriano terrane. This interpretation is supported by: (1) preserved geometrical relationships; (2) the existence of high pressure metamorphic relics in the neighbourhood of the suture; (3) evidence of ophiolite obduction; (4) the development of a foreland basin onto the overridden block, the successive depocenters of which migrated towards the north, presumably as a consequence of the migration of deformation in the same direction; and (5) evolution of the geometrical characteristics of the structures developed during this second Cadomian phase, which are antithetic with respect to the subduction zone in the upper block and synthetic in the overridden block. In the latter, they evolve from flat-lying near the suture, through north-verging, to upright in northern positions.

Every tectonic activity associated with the Cadomian orogeny seems to have ceased by lowermost Cambrian times, as is demonstrated by the unconformable relationships with basal Cambrian clastic formations (partly red beds) marking the initiation of a new rifting event. The axis of maximum extension during this new event was located south and west of the Ossa Morena Zone and resulted in the break-up of the continent with newly accreted terranes that had been formed during the Cadomian collision. The Palaeozoic Iberian terrane was formed in this process including pieces of the various amalgamated Precambrian terranes, the suture within which has been repeatedly reactivated in Palaeozoic times—as a normal fault with a strike-slip component during the Lower Palaeozoic rift process, and as a wrench and thrust fault during Variscan orogenesis (Ribeiro et al., 1989). Collectively all these tectonic processes (Precambrian and Palaeozoic) have resulted in the formation of a very complex belt which is currently known in the literature as the Badajoz-Córdoba shear zone (see Figure 7.18).

Acknowledgements

Many of the ideas expressed in this paper have been refined through long, constructive discussions with O. Apalategui, L. A. Cueto, V. Monteserin, J. Munhá, J. T. Oliveira and A. Ribeiro. The providing of original pictures by F. Moreno is also kindly acknowledged as is the efficient professional assistance by P. Asensio (typing) and J. Vallejo (artwork drawing).

References

Aguayo, J. (1985) Rocas ultramáficas en el sector de Calzadilla de los Barros (Badajoz). *Tesis Licenciatura Univ. Bilbao*, 1–110.

Andrade, A. A. S., Borges, F. S., Marques, M. M., Noronha, F. and Pinto, M. S. (1983) Contribuição para o conhecimiento da faixa metamórfica da Foz do Douro (Nota prévia). *I Congr. Natl. Geol. Portugal*, Aveiro (*Abstract*).

Apalategui, O. and Higueras, P. (1983) *Mapa Geológico de España*, escala 1:50.000 (2a. serie). Sheet no. 855: Usagre. *I.G.M.E.*

Apalategui, O. and Quesada, C. (1987) Transversal geológica zona Ossa-Morena. *Field Excursion Guide Book*, I.G.M.E., 1–98.

Apalategui, O., Borrero, J. D. and Higueras, P. (1985) Division en grupos de rocas en Ossa-Morena oriental. *Temas Geol. Min.* **8**, 73–80.

Apalategui, O., Garrote, A., Roldan, F. J. and Sanchez, R. (1986) *Mapa Geológico de España*, escala 1:50.000 (2a. serie). Sheet no. 879: Peñarroya-Pueblonuevo. *I.G.M.E.*

Arriola, A., Garrote, A., Eguiluz, L., Sanchez, R. and Portero, J. M. (1983) *Mapa Geológico de España*, escala 1:50.000 (2a. serie). Sheet no. 876: Fuente de Cantos. *I.G.M.E.*

Arriola, A., Cueto, L. A., Fernandez-Carrasco, J. and Garrote, A. (1984) Serpentinitas y mineralizaciones de cromo asociadas en el Proterozoico Superior de Ossa-Morena. *Cuad. Laboratorio Xeolóxico Laxe* **8**, 137–146.

Bischoff, L., Wildberg, H. and Baumann, A. (1986) Uranium/Lead ages of zircons from gneisses of the Sistema Central, Central Spain. *Int. Conf. on Iberian Terranes and their Regional Correlation*, IGCP Project No. 233, Annu. Meet., Oviedo (Spain). *Abstract Volume*, **39**.

Chacón, J. (1979) Estudio geológico del sector central del anticlinorio Portoalegre-Badajoz-Córdoba (SW del Macizo Ibérico). *Tesis Doctoral Univ. Granada*, 1–721.

Chacón, J., Fernandez-Carrasco, J., Mitrofanov, F. and Timofeev, B. V. (1984) Primeras dataciones microfitopaleontologicas en el sector de Valverde de Burguillos-Jerez de los Caballeros (anticlinorio Olivenza-Monesterio). *Cuad. Laboratorio Xeolóxico Laxe* **8**, 211–220.

Coney, P., Jones, D. L. and Monger, J. W. H. (1980) Cordilleran Suspect Terranes. *Nature (London)* **288**, 329–333.

Cueto, L. A., Eguiluz, L., Llamas, J. F. and Quesada, C. (1983) La granodiorita de Pallares, un intrusivo precámbrico en la alineación Olivenza-Monesterio (Zona de Ossa-Morena). *Comun. Serv. Geol. Port.* **69** (2), 219–226.

Delgado, M. (1971) Esquema geológico de la hoja num. 878 de Azuaga (Badajoz). *Bol. Geol. Min.* **82**, 277–286.

Eguiluz, L. (1987) Petrogénesis de rocas ígneas y metamórficas en el antiforme Burguillos-Monesterio. Macizo Ibérico Meridional. *Tesis Doctoral Univ. Bilbao*, 1–694.

Fernández, J., Coullaut, J. L., Eguiluz, L. and Garrote, A. (1983) *Mapa Geológico de España*, escala 1:50.000 (2ª serie). Sheet nº 897: Monesterio. *I.G.M.E.*

Gabaldon, V., Hernandez, J., Lorenzo, S., Picart, J., Santamaria, J. and Sole, J. (1989) Sedimentary facies and stratigraphy of phosphate rocks in the Valdelacasa structure (Precambrian–Cambrian). Central Iberian Zone (Spain). In *Phosphate Deposits of the World, Vol. 2: World Phosphate Resources*, eds. Notholt, A. J. G., Sheldon, R. P. and Davidson, D. F., Cambridge Univ. Press (in press).

Galindo, C., Casquet, C., Portugal Ferreira, M. and Macedo, C. A. P. (1987) Geocronología del Complejo Plutónico Táliga-Barcarrota (CPTB) (Badajoz, España). In *Geología de los Granitoides y Rocas Asociadas del Macizo Hespérico*, eds. Bea, F., Carnicero, A., Gonzalo, J. C., López Plaza, M. and Rodríguez Alonso, M. D., Ed. Rueda, Madrid, 385–392.

Garcia-Casquero, J. L. (1989) Continental trondhjemites in displaced terranes of the Oporto-Portalegre-Badajoz-Córdoba Belt (Hesperian Massif). *Geol. Rundsch.* (in press).

Garrote, A. (1976) Asociaciones minerales del núcleo metamórfico de Sierra Albarrana (prov. de Córdoba). *Mem. Not. Publ. Mus. Lab. Mineral. Geol. Univ. Coimbra Cent. Estud. Geol.* **82**, 17–40.

Garrote, A., Delgado, M. and Contreras, M. C. (1989). *Mapa Geológico de España*, escala 1:50.000 (2a. serie), Sheet no. 900: La Cardenchosa. *I.G.M.E.* (in press).

Iglesias, M., Ribeiro, M. L. and Ribeiro, A. (1983) La interpretación aloctonista de la estructura del noroeste peninsular. In *Geologia de España, Libro Jubilar J.M. Ríos*, I.G.M.E. **1**, 459–466.

Julivert, M., Fontbote, J. M., Ribeiro, A. and Conde, L. (1974) *Mapa Tectónico de la Península Ibérica y Baleares*. Escala 1:1.000.000. *I.G.M.E.*

Kuijper, R. P. (1979) U–Pb systematics and the petrogenetic evolution of infracrustal rocks in the Palaeozoic basement of Western Galicia, northwest Spain. *Verh, ZWO Lab. Isot. Geol.* **5**, 1–101.

Kuijper, R. P., Priem, H. N. A. and Den Tex, E. (1982) Late Archaean–Early Proterozoic source areas of zircons in rocks from the Palaeozoic orogen in Western Galicia. *Precambrian Res.* **19** (1), 1–29.

Lancelot, J. R., Allegret, A. and Iglesias, M. (1985) Outline of Upper Precambrian and Lower Palaeozoic evolution of the Iberian Peninsula according to U–Pb dating of zircons. *Earth Planet. Sci. Lett.* **74**, 325–337.

Liñán, E. (1978) Bioestratigrafía de la Sierra de Córdoba. *Tesis Doctoral Univ. Granada* **191**, 1–212.

Liñán, E. and Schmitt, M. (1980) Microfósiles de las calizas precámbricas de Córdoba (España). *Temas Geol. Min.* **4**, 171–194.

Lotze, F. (1945) Zur gliederung der Varisciden der Iberischen Meseta. *Geoteck. Forsch.* **6**, 78–92.

Macpherson, J. (1881) Apuntes petrográficos de Galicia. *An. Soc. Esp. Hist. Natural* **10**, 49–87.

Mata, J. and Munha, J. (1986) Geodynamic significance of high grade metamorphic rocks from Degolados-Campo Maior (Tomar-Badajoz-Córdoba Shear Zone). *Maleo* **2** (13), 28.

Moreno, F. (1977) Estudio geológico de los Montes de Toledo occidentales. *Tesis Doctoral Univ. Complutense*, Madrid, 1–483.

Munha, J., Oliveira, J. T., Ribeiro, A., Oliveira, V., Quesada, C. and Kerrich, R. (1986) Beja-Acebuches Ophiolite: characterization and geodynamic significance. *Maleo* **2** (13), 31.

Nägler, TH. F., Gebauer, D., Schäfer, H. J. and Von Quadt, A. (1988) Sm–Nd and Pb-isotope geochemistry of Palaeozoic sediments of the Almadén and Guadalmez synclines (Central Iberian Zone, Spain). In *Cinturones Orogénicos*, II Congreso Geológico de España, *Symposia volume*, 45–50.

Nägler, TH. F., Gebauer, D. and Schäfer, H. J. (1989) Nd-, Sr- and Pb isotope geochemistry of two pre-Permian sedimentary profiles from the Spanish Meseta: Evidence of both crustal recycling and crustal growth. *Terra Abstr.* **1** (1), 336.

Palacios, T. (1986) Acritarcos Proterozoico Superior-Vendiense en el Complejo Esquisto-Grauváquico de Extremadura. *Tesis Doctoral Univ. Extremadura*, 1–530.

Parga-Pondal, I. (1956) Nota explicativa del mapa geológico de la parte NO de la provincia de La Coruña. *Leidse Geol. Meded.* **21**, 467–484.

Parga-Pondal, I. (1960) Observación, interpretación y problemas geológicos de Galicia. *Notas Comun. IGME* **59**, 333–358.

Parga-Pondal, I. (1963) *Mapa Petrográfico-Estructural de Galicia* E. 1: 400.000. *I.G.M.E.*

Pascual, E. (1981) Investigaciones geológicas en el sector Córdoba-Villaviciosa de Córdoba (Sector Central de Sierra Morena). *Tesis Doctoral Univ. Granada*, 1–521.

Pérez, F. (1979) Geología de la zona Ossa-Morena al Norte de Córdoba (Pozoblanco-Belmez-Villaviciosa de Córdoba). *Tesis Doctoral Univ. Granada* **281**, 1–340.

Pieren, A. P., Pineda, A. and Herranz, P. (1987) Discordancia intra-Alcudiense en el anticlinal de Agudo (Ciudad Real-Badajoz). *Geogaceta* **2**, 26–29.

Quesada, C. (1975) Geología de un sector de la parte central del anticlinorio Olivenza-Monesterio. Alrededores de Monesterio (Badajoz). *Tesis Licenciatura Univ. Granada*, 1–128.

Quesada, C. (1990) Precambrian successions in southwest Iberia: their relationship to Cadomian orogenic events. In *The Cadomian Orogeny*, eds. D'Lemos, R. S., Strachan, R. A. and Topley, C. G., *Spec. Publ. Geol. Soc. London* **51**, 353–362.

Quesada, C. and Munha, J. (1989) Metamorphism within the Ossa-Morena Zone. In *Pre-Mesozoic Geology of the Iberian Peninsula*, eds. Dallmeyer, R. D. and Martínez, E., (in press).

Quesada, C., Florido, P., Gumiel, P. and Osborne, J. (1987) *Mapa geológico-minero de Extremadura. Junta de Extremadura*, 1–131.

Ribeiro, A. (1983) Los complejos de Bragança y Morais. In *Geología de España, Libro Jubilar J.M. Ríos*, I.G.M.E., **1**, 450–454.

Ribeiro, A., Dias, R., Pereira, E., Merino, H., Sobre Borges, F., Noronha, F. and Marques, M. (1987a) Guide-book for the Miranda do Douro-Porto Excursion. *Conference on Deformation and Plate Tectonics*, Gijón (Spain), 1–50.

Ribeiro, A., Quesada, C. and Dallmeyer, R. D. (1987b) Tectonostratigraphic terranes and the geodynamic evolution of the Iberian Variscan Fold Belt. *Conference on Deformation and Plate Tectonics*, Gijón (Spain), *Abstract Volume*, 60–61.

Ribeiro, A., Quesada, C. and Dallmeyer, R. D. (1989) Geodynamic evolution of the Iberian Massif. In *Pre-Mesozoic Geology of the Iberian Peninsula*, eds. Dallmeyer, R. D. and Martínez, E., (in press).

Ries, A. C. and Shackleton, R. M. (1971) Catazonal complexes of Northwest Spain and North Portugal, remnants of a Hercynian thrust plate. *Nature (London)* **234** (47), 65–68.

Sánchez Carretero, R., Eguiluz, L., Pascual, E. and Carracedo, M. (1989) Igneous rocks of the Ossa-Morena Zone. In *Pre-Mesozoic Geology of the Iberian Peninsula*, eds. Dallmeyer, R. D. and Martínez, E., (in press).

Schäfer, H. J., Gebauer, D., Nägler, TH. F. and Von Quadt, A. (1988) U–Pb zircon and Sm–Nd studies of various rock-types of the Ossa-Morena Zone (Southwest Spain). In *Cinturones Orogenicos*, II Congreso Geológico de España, Granada, *Symposia volume*, 51–57.

Schäfer, H. J., Gebauer, D. and Nägler, TH. F. (1989) Pan-African and Caledonian ages in the Ossa-Morena Zone (southwest Spain): A U–Pb and Sm–Nd study. *Terra Abstr.* **1** (1), 350–351.

Schulz, G. (1835) Descripción geognóstica del Reino de Galicia. *Gráficas Reunidas*, S.A., Madrid, 1–138.

Vilas, L., Garcia, J. F., San Jose, M. A., Pieren, A. P., Pelaez, J. R., Perejon, A. and Herranz, P. (1987) Episodios sedimentarios en el Alcudiense Superior (Proterozoico) y su tránsito al Cámbrico en la zona centro meridional del Macizo Ibérico. *Geogaceta* **2**, 43–45.

8 The West African orogens and Circum-Atlantic correlatives

R. D. DALLMEYER

8.1 Introduction

Most late Palaeozoic continental reconstructions place western Africa adjacent to southeastern North America (e.g., Van der Voo et al., 1976; Scotese et al., 1979; Lefort, 1980; Klitgord and Schouten, 1981; Pilger, 1981; Van der Voo, 1983; Ross et al., 1986; Rowley et al., 1986; Ross and Scotese, 1989). These suggest potential tectono-thermal linkages between the central–southern Appalachian orogen and the Mauritanide, Bassaride and Rokelide orogens of West Africa. All of these areas are dominated by variably allochthonous sequences which when traced tectonostratigraphically upward become increasingly more exotic relative to the structurally underlying West African sequences. These exotic lithostratigraphic terranes represent successions which could have formed: (1) outboard of the Laurentian or Gondwanan during middle–late Palaeozoic closure of the Iapetus oceanic tract; (2) in early–middle Palaeozoic settings removed from either Laurentia or Gondwana and accreted either prior to or during amalgamation of Pangea; and/or (3) Laurentian or Gondwanan continental fragments stranded during Mesozoic rifting associated with opening of the present Atlantic ocean.

Results of recent collaborative field and geochronological work in West Africa and the southern Appalachian orogen have helped to resolve the origin of many of these exotic terranes and establish their accretionary chronology. These results are briefly summarized in this contribution.

8.2 Mauritanide, Bassaride, and Rokelide orogens

The Mauritanide, Bassaride, and Rokelide orogens occur along the western edge of the West African Craton (Figure 8.1). The orogens are comprised of four contrasting tectonostratigraphic sequences, including: (1) crystalline basement of the West African Shield (Archaean–middle Proterozoic); (2) autochthonous/parautochthonous sedimentary cover sequences (upper Proterozoic–late Palaeozoic); (3) a series of infrastructural allochthonous com-

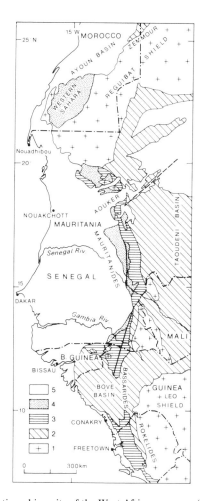

Figure 8.1   Tectonostratigraphic units of the West African orogens (adapted from Lécorché *et al.*, 1990): 1 crystalline basement of the West African Shield (Archaean–middle Proterozoic); 2 autochthonous/parautochthonous sedimentary cover sequences (upper Proterozoic–late Palaeozoic); 3 infrastructural allochthonous complexes dominated by upper Proterozoic–lower Palaeozoic (Pan-African) orogenic terranes; 4 suprastructural exotic allochthonous complexes emplaced during late Palaeozoic (Hercynian) orogeny; 5 Mesozoic and younger sedimentary sequences.

plexes largely comprised of upper Proterozoic–lower Palaeozoic (Pan-African) orogenic terranes; and (4) allochthonous, exotic suprastructural crystalline complexes emplaced during the late Palaeozoic (Hercynian) orogeny. Mesozoic and younger sedimentary sequences unconformably overlie the West African orogens in a series of western coastal basins. Geological reviews of the West African orogens have been provided by Allen (1967, 1969),

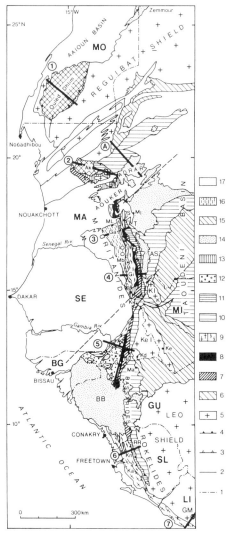

Figure 8.2 Generalized geologic map of the West African orogens (adapted from Lécorché *et al.*, 1989): 1 political boundaries; 2 faults; 3 Pan-African thrust; 4 late Palaeozoic thrust; 5 undifferentiated crystalline basement (West African Shield); 6 Supergroup 1; 7–9 infrastructural allochthonous complexes (7 Unit A, 8 Unit B, 9 Unit C); 10 Supergroup 2; 11 S2a; 12 S2b; 13 undifferentiated S2b + S3; 14 Supergroup 3; 15, 16 suprastructural allochthonous complexes (15 internal nappes of Lécorché *et al.*(1983), 16 westernmost Hercynian(?) formations); 17 Mesozoic and younger cover. Abbreviations: Political: BG = Guinea Bissau, GU = Guinea, LI = Liberia, MA = Mauritania, ML = Mali, MO = Morocco, SE = Senegal, SL = Sierra Leone. Regions: AS = Assaba, BB = Bové Basin, GM = Gibi Mountain, Kel = Kéniéba inlier, KI = Kayes Inlier, T = Taganet. Localities: A = Agoualilet, Ak = Akjoujt, B = Bakel, Bn = Bou Naga window, E = el Aoueija, F = Farkâka, G = Guingan, Ga = Gaoua, Gi = Guidimakha, K = Kayes, Ka = Kasila, Kd = Dedougou, Ke = Kéniéba, Ki = Kidra, Kn = Kenema, Ko = Koulountou, Ma = Mali, Mb = Mbout, Mj = Mejeira, Ml = Magta Lahjar, S = Sangarafa, T = Termessé, Y = Youkoukoun. Lines of the cross-sections shown in Figure 8.3(A) and Figure 8.4(1–7) are located.

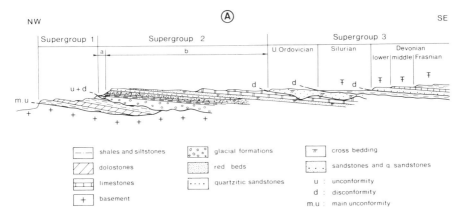

Figure 8.3 Generalized tectonostratigraphic cross-section of the Mauritanian Adrar (line A, Figure 8.2: adapted from Lécorché *et al.*, 1989).

Williams (1978), Lécorché (1980), Lécorché *et al.* (1983), Roussel *et al.* (1984), Villeneuve and Dallmeyer (1987) and Lécorché *et al.* (1989, 1990).

8.2.1 *Geologic setting*

Crystalline basement for the various West African orogens is exposed within the Reguibat and Leo Shields and in the Kayes and Kenieba (Eastern Senegal) inliers (Bessoles, 1977; Williams, 1978). Basement sequences include Liberian (*c.* 2700 Ma) high grade gneisses and associated catazonal granitic plutons together with less metamorphosed volcanic and sedimentary rocks of Eburnian age (*c.* 2000–1800 Ma). Widespread granitic plutonism occurred throughout the basement at *c.* 1800 Ma.

An extensive late Proterozoic sedimentary sequence (Supergroup 1 of Lécorché *et al.*, 1983) unconformably overlies the basement (Figures 8.1–8.3). This appears to range in age between *c.* 1100 and 700 Ma (Bassot *et al.*, 1963; Clauer, 1976; Clauer *et al.*, 1982), and is dominated by shallow marine rocks in the north and continental clastic units in the south (S1, Table 8.1). North of the Rokelide orogen, an infrastructural allochthonous sequence of variably deformed and metamorphosed volcanic and metasedimentary rocks (Figures 8.2 and 8.4) comprise a late Proterozoic rift sequence which could, in part, be time-correlative with upper portions of Supergroup 1 (Villeneuve, 1980). The infrastructural units include volcanic–volcaniclastic rocks (with both tholeiitic and alkaline basalts: Dupont *et al.*, 1984; Rémy, 1987), serpentinite and intercalated marine sequences (chert, jasper, and graywacke). It is represented by two distinct, internally imbricated structural units: (1) a slightly meta-

Table 8.1 Tectonostratigraphic correlations in the Mauritanide, Basseride and Rokelide orogens, West Africa (from Lécorché et al., 1989).

Orogens	Country	Regions	Regional Synthesis recent - (older)	Westernmost/Uppermost Formations	Internal Formations	Axial Formations	External Formations	Immediate Foreland
MAURITANIDES	Morocco	Southern Provinces	Bronner, Marchand and Sougy 1983	Ouled Dhlim nappes	Unknown	Unknown	Unknown	S.3 / basement
MAURITANIDES	Mauritania	N-Western	Lécorché, 1980 (Marcelin, 1975)	Internal nappes (are related tholeiites + basement) / Quartzite nappe (S.3, S.2b?)		Agouallet (U.A ?)	External nappes (S.2a, U.A) Window (S.2a, basement)	S.3 S.2b S.2a basement
MAURITANIDES	Mauritania	Central	Dia, 1984	Tingarach Sangarafa Gaoua (S.3, S.2b) / Magta Lahjar	El Khneikat Kelbé (Calc-alk)	Farkaka (U.B.) El Aoueija (U.A)	Djonaba (S.2a)	S.3 S.2b S.2a ?
MAURITANIDES	Senegal	Southern	(Chiron, 1973) (Lille, 1967) Le Page, 1983	Oua-Oua	Mbout Guidimakha (c-alk)	Diala-Bouanzé (UB) Hamdallaye (UA)	Mbaïou (S.2-3)	S.3 S.2b S.2a basement
MAURITANIDES	Guinea	N-Eastern	(Bassot, 1966)	Oundou Baba	Debi (Calc-alk)	Gabou (U.A)	Kidira (S.2a)	
BASSARIDES	Senegal	Central S-Eastern	Villeneuve, 1984	Youkounkoun (S.2b) folds Bové Basin (S.3)	Damantan / Koudountou (Calc-alk)	Guingan (U.B)	Mali (S.2a)	S.2a / S.1 basement
BASSARIDES	Guinea	Northern Southern	(Allen, 1969) Williams, 1978	Taban (S.2b)	Forecariah (basement) Kasila (basement)	Termesse (U.A)	Kolenté (S.2a)	S.2a basement
ROKELIDES	Sierra Leone		Thorman, 1976	unknown	Coastal zone (basement)	unknown	Kenema (basement) Maramba (basement or U.A) Rokel River (S.2a)	
ROKELIDES	Liberia						Gibi Mountain (S.2a)	basement

morphosed, tholeiitic to alkaline eastern volcanic sequence (Unit A, Table 8.1); and (2) a higher grade (amphibolite facies) metavolcanic and meta-sedimentary western sequence (Unit B, Table 8.1). Within the central Mauritanides, the infrastructural western sequence (Farkâka Association) includes continental supracrustal components and intra-plate, rift-related metabasalts (Rémy, 1987). Rocks within the western sequence have been

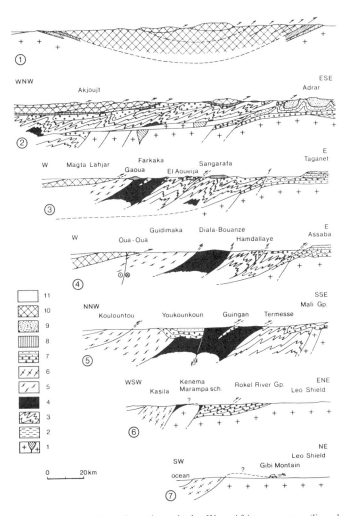

Figure 8.4 Schematic tectonic sections through the West African orogens (lines located on Figure 8.2: adapted from Lécorché *et al.*, 1989): 1 undifferentiated basement with *c.* 680 Ma peralkaline intrusions; 2 Supergroup 1; 3–5 infrastructural allochthonous complexes (3 Unit A, 4 Unit B, 5 Unit C); 6 exotic basement (within infrastructural and/or suprastructural allochthonous complexes); 7 S2a (circles) or S2b (dots); 8 undifferentiated S2B + S3; 9 Supergroup 3; 10 suprastructural allochthonous complexes; 11 Mesozoic and younger cover.

interpreted to have formed, in part, along the margin of a continental block (Senegal microplate of Lefort, 1980) which rifted from the West African Craton in the late Proterozoic (Lécorché et al., 1989). Eastern, lower grade infrastructural sequences probably developed along the western margin of the West African Craton during during this rifting. Blanc (1986) reported U-Pb zircon crystallization ages of c. 680 Ma for alkaline and peralkaline felsic intrusions believed to have been emplaced during this rifting event.

Westermost infrastructural units of the West African orogens are dominated by variably retrogressed, mylonitic gneissic rocks (in part, Unit C, Table 8.1). In the Mauritanides and Bassarides these are host to a generally calc-alkaline, variably deformed and metamorphosed igneous sequence which includes felsic volcaniclastic units and associated comagmatic granitic plutons. Rb-Sr whole-rock isochron and $^{40}Ar/^{39}Ar$ mineral ages suggest that the igneous sequence developed at c. 685–675 Ma (Bassot and Caen-Vachette, 1983; Dallmeyer and Villeneuve, 1987).

An epicontinental to marine clastic sequence (Supergroup 2 of Lécorché et al., 1983) unconformably overlies Supergroup 1 and various pre-deformed and metamorphosed infrastructural units of the West African orogens (Figures 8.1–8.4). Basal portions of this sequence (S2a, Table 8.1) are marked by a distinctive upper Proterozoic tillite which is often associated with baritic carbonate, chert, and stromatolitic dolostone. Culver et al. (1988) have described early Cambrian microfossils (Aldanella) from conformably overlying flyschoid shales in northern Guinea. Parautochthonous, western external structural units of the West African orogens contain deformed and slightly metamorphosed sequences which are correlative with lower portions of Supergroup 2 successions in the foreland. These are locally thrust eastward over deformed but nonmetamorphosed foreland rocks. Traced westward, the external sequences become progressively more allochthonous and are imbricated with infrastructural tectonic units (particularly Units A and B).

Upper portions of Supergroup 2 (S2b, Table 8.1) are most extensively developed in the Mauritanide foreland where they are represented by cross-bedded, feldspathic red sandstones which locally unconformably overlie lower portions of Supergroup 2. The red sandstones are conformably overlain by sequences of cleanly washed, scolithic white sandstones. Basal units contain inarticulate brachiopod faunas which span the Cambrian–Ordovician boundary (Legrand, 1969). In the Bassaride and Rokelide orogens, coarse, molassic red sandstones unconformably overlie various deformed and metamorphosed structural units; scolithic white sandstones are absent.

Sequences related to a widespread Upper Ordovician glaciation (Deynoux, 1978) mark the base of a third foreland succession in the Mauritanide foreland (Supergroup 3 of Lécorché et al., 1983): S3, Figures 8.1–8.4 and Table 8.1). The basal erosional disconformity is locally complicated by an angular unconformity of uncertain regional significance (Dia et al., 1969; Dia, 1984). In the Bassaride and northern Rokelide orogens, possibly related, cross-bedded quartzitic sandstones occur in basal units exposed in the Bové Basin (Figures 8.1–8.4). These unconformably overlie either molassic red sandstones (S2b), pre-deformed late Proterozoic early Cambrian flyschoid shales (S2a), or pre-deformed and metamorphosed, infrastructural structural units (axial units A and B, Table 8.1). In Mauritania, basal sequences of Supergroup 3 are conformably overlain by Early Devonian sandstone and shale, Middle Devonian limestone and Upper Devonian siltstone and shale which range up to Frasnian in age (Dia, 1984). In the Bové Basin, the basal formations are conformably overlain by Silurian shales which may be traced continuously upward into Devonian sandstones and shales which range into the Fammenian. No Carboniferous sequences have been reported from the West African orogens.

Metasedimentary rocks within several suprastructural West African allochthonous units have uncertain foreland correlations. These include a sequence of very low grade, deformed sandstones which contain Lower Devonian faunas and structurally overlie pre-deformed, infrastructural high grade units west of Mejeria (Figures 8.2 and 8.4). The low grade metasedimentary sequence may be correlated with either upper portions of Supergroup 2 or with Supergroup 3. Similar uncertain correlations apply to an extensive, suprastructural mylonitic quartzite nappe exposed near Akjoujt in northwestern Mauritania (Figures 8.2 and 8.4). The mylonitic quartzite nappe is structurally overlain by an internally imbricated allochthonous suprastructural complex (the internal nappes of Lécorché, 1980; Lécorché et al., 1983). Lower tectonic levels of the nappe complex are represented by alternating structural units comprised of low grade metapelitic rocks, mafic volcanic sequences, and iron quartzites. The occurrence of K-poor, undifferentiated tholeiitic basalts within the metavolcanic intervals suggest an island arc or back arc origin (Ba Gatta, 1982; Kessler, 1986).

Intermediate tectonic levels of the suprastructural nappe complex include structural units of gneissic rocks with subordinate, interlayered amphibolite (Lécorché, 1980). Hornblende and muscovite from these units display internally discordant $^{40}Ar/^{39}Ar$ age spectra (Dallmeyer and Lécorché, 1990a) which reflect variable late Palaeozoic thermal rejuvenation of argon systems which record initial post-metamorphic cooling ages between 2400 and 1800 Ma. Upper tectonic levels of the nappe complex are largely represented by variably mylonitic and extensively retrogressed gneissic rocks (Giraudon and Sougy, 1963; Marcelin, 1975). Muscovite developed along fluxion fabrics in mylonitic metagranite record 310–300 Ma $^{40}Ar/^{39}Ar$ plateau ages (Dallmeyer and Lécorché, 1990a). A similar, internally imbricated, suprastructural nappe complex also occurs as an extensive klippe northwest of the Reguibat shield (Figures 8.2 and 8.4; Sougy, 1962; Bronner et al., 1983). A variety of variably metamorphosed and deformed rocks are included within this nappe complex and they may, in part, be structurally correlative with the Akjoujt suprastructural nappe complex.

8.2.2   *Tectono-thermal history*

Field relationships and available geochronological controls suggest that the effects of three distinct tectono-thermal events are locally recorded in the various West African orogens. In the Bassarides, volcanic and associated sedimentary units within the infrastructural late Proterozoic rift sequence (Units A and B; Termesssé and Guingan Groups of Villeneuve, 1984) were deformed and variably metamorphosed (greenschist to amphibolite facies) prior to deposition of the unconformably overlying late Proterozoic tillite (S2a: base of Mali Group; Villeneuve, 1984). Dallmeyer and Villeneuve (1987) reported 660–650 Ma ^{40}Ar/^{39}Ar plateau ages for muscovite within penetratively cleaved metasedimentary rocks of the rift sequence in southern Senegal (Guingan Group; Unit B, Table 8.1), and interpreted these to closely date a Pan-African I phase of tectono-thermal activity. Similar 650–620 Ma mineral ages (K-Ar, ^{40}Ar/^{39}Ar and Rb-Sr) were presented by Lille (1967) and Dallmeyer and Lécorché (1990b) for variably mylonitic, calc-alkaline meta-granites within the infrastructural Guidimakha allochthon (Unit C, Table 8.1) in the southern Mauritanides (north of Bakel, Figure 8.2; section 4, Figure 8.4). In addition, 660–650 Ma ^{40}Ar/^{29}Ar plateau ages were reported by Dallmeyer and Lécorché (1989a) for hornblende within intermediate tectonic levels of infrastructural allochthonous units (amphibolite within the Fakâkar association: UB, Table 8.1) in central portions of the Mauritanides (southeast of Magta Lahjar, Figure 8.2: section 3, Figure 8.4). The available geochronological controls, together with a widespread sedimentological expression of tectonic instability in the West African Craton at *c.* 650 Ma (Bronner *et al.*, 1980) combine to suggest that Pan-African I orogenesis was of regional significance.

In the Bassarides of southern Senegal (section 5, Figure 8.4) the late Proterozoic–early Cambrian Mali Group (S2a) was locally folded prior to deposition of unconformably overlying (S2b) molasse sequences (Youkoukoun and Taban Groups; S2b; Villeneuve, 1984). This tectono-thermal event appears to have increased in intensity westward where calc-alkaline igneous rocks within infrastructural allochthons record up to greenschist facies metamorphic assemblages (Villeneuve, 1984). Dallmeyer and Villeneuve (1987) reported *c.* 550 Ma ^{40}Ar/^{39}Ar plateau ages for muscovite within penetratively cleaved felsic metavolcanic rocks from this area (Unit C: Simenti Group of Villeneuve, 1984), and suggested that these closely date a Pan-African II phase of tectono-thermal activity. The effects of Pan-African II tectono-thermal activity are penetratively recorded throughout the Rokelide orogen. In Sierra Leone (section 6, Figure 8.4) components of a western gneissic succession are imbricated with both cover sequences (S2a: Rokel River Group which is, in part, equivalent to the Mali Group; Villeneuve, 1984) and penetratively mylonitic and thoroughly retrogressed structural units of West African Shield basement (Allen, 1967, 1969; Williams, 1978). Cover sequences are absent in Liberia (section 7, Figure 8.4) where the

Rokelides are characterized by imbrication of the western gneissic sequence and West African Shield basement (Thorman, 1976). In northern portions of the Rokelide orogen (southern Guinea and northern Sierra Leone), hornblende records K-Ar and ^{40}Ar/^{39}Ar plateau ages ranging between c. 580 and 550 Ma (Allen et al., 1967; Beckinsale et al., 1981; Dallmeyer, 1989a). The hornblende ages are interpreted to date post-metamorphic cooling through temperatures required for intracrystalline retention of argon. Post-meta-morphic cooling following Pan-African II tectono-thermal activity appears to have been younger in the southern Rokelides where hornblende yields K-Ar dates between c. 530 and 485 Ma (Hurley et al., 1971; Hedge et al., 1975). This probably reflects an increased diachronism between attainment of maximum Pan-African II thermal conditions and subsequent cooling through hornblende argon retention temperatures. This is interpreted to reflect exposure of increasingly deeper structural levels southward in the Rokelide orogen (consistent with preservation of cover sequences only in northern areas). Pan-African I orogenesis has not been documented within the Rokelide orogen. It is uncertain whether this reflects an initial absence of a Pan-African I record, or its complete removal during the penetrative Pan-African II tectono-thermal activity.

Late Palaeozoic folding and thrusting affects foreland sequences (which range into the Late Devonian) along the length of the Mauritanide orogen. Limited geochronological data suggest tectono-thermal activity of similar age is locally recorded within western infrastructural and suprastructural tectonic units. Lécorché and Clauer (1983) reported c. 300 Ma K-Ar ages for < 2 μm illite size fractions within samples collected adjacent to the thrust at the base of the suprastructural nappe complex (near Akjoujt. Figures 8.1 and 8.2). In the same area, Dallmeyer and Lécorché (1990a) reported 310–300 Ma ^{40}Ar/^{39}Ar plateau ages for dynamically recrystallized muscovite within mylonitic metagranite from structurally higher suprastructural units. The late Palaeozoic thermal overprint appears to have been less intense within intermediate tectonic levels of the suprastructural nappe complex. These sequences yield 2400–1800 Ma muscovite and hornblende ^{40}Ar/^{39}Ar plateau ages which display no record of a Pan-African tectono-thermal overprint. Dallmeyer and Lécorché (1989b) suggested that these sequences represent late Palaeozoic structural units which were derived from a basement terrane (Senegal microplate) presently concealed beneath the Senegal-Mauritania coastal basin.

In the Central Mauritanides, Dallmeyer and Lécorché (1989) reported ^{40}Ar/^{39}Ar age spectra for muscovite separated from samples from the suprastructural Gâoua quartzite nappe. These become increasingly more discordant structurally downward in the nappe. Dallmeyer and Lecorche (1989) interpreted this variation to reflect the increasing effects of a c. 310–300 Ma thermal overprint (associated with final nappe emplacement) on intracrystalline muscovite argon systems which had earlier cooled below closure temperatures at c. 580 Ma. Complete rejuvenation of muscovite argon systems at c. 300 Ma was effected only in westernmost infrastructural units (cleaved and variably metamorphosed calc-alkaline metavolcanic sequences) exposed west of Mejeria (Figure 8.2; Dallmeyer and Lécorché, 1989). Together, the K-Ar and ^{40}Ar/^{39}Ar results clearly indicate that late Palaeozoic thrusting and metamorphism affected both suprastructural and previously imbricated, westernmost infrastructural allochthons within the Mauritanide orogen. Deformation of similar age is likely observed throughout the northern Bassarides and probably extends south to the Bové basin where Late Ordovician–Devonian sedimentary successions are folded with local development of a fracture cleavage (Villeneuve, 1984). Dallmeyer and Villeneuve (1987) reported ^{40}Ar/^{39}Ar plateau ages of c. 280–270 Ma for dynamically recrystallized muscovite within fluxion fabrics in mylonitic calc-alkaline granites in northeastern Senegal west of Kedougou (Figures 8.1 and 8.2). These suggest that significant late Palaeozoic tectono-thermal activity is at least locally recorded in northwesternmost portions of the Bassarides. These effects do not extend into central Guinea (Villeneuve and Dallmeyer, 1987), and there is no record of late Palaeozoic tectono-thermal activity in the Rokelide orogen.

8.2.3 Geophysical controls on tectonic models

The Mauritanide orogen is bordered on the east by the West African Craton and on the west by a thick section of Mesozoic and younger sedimentary sequences within the Senegal-Mauritania coastal basin (Figures 8.1 and 8.2). The regional Bouguer gravity anomaly pattern suggests that the orogen extends westward beneath the basin (Lécorché et al., 1983; Roussel et al., 1984). Three distinct anomaly domains have been outlined: (1) a prominent, nearly continuous, NNW–SSE trending belt of positive anomalies known as the Mauritanian anomaly, which, although paralleling the axis of the exposed portions of the orogen, is displaced slightly westward; (2) a broad regional negative anomaly east of the Mauritanian anomaly which is characterized by NE–SW gravity trends; and (3) a generally positive anomaly west of the Mauritanian anomaly. Lécorché et al. (1983) and Roussel et al. (1984) interpreted the regional gravity pattern east of the Mauritanian anomaly to largely reflect the signature of Precambrian basement beneath the West African Craton. This distinctive signature may be traced westward beneath exposed portions of the orogen, clearly reflecting its allochthonous nature. The strong positive Mauritanian anomaly has been interpreted as an asymmetric mantle ridge oversteepened eastward which has a crest at a depth of c. 15 km. East of longitude 15°30'W the gravity character over the Mesozoic basin is very similar to that defined by Precambrian crystalline rocks beneath the West African Craton. This led Ponsard et al. (1982). Lécorché et al. (1983), Ponsard (1984) and Roussel et al. (1984), to propose that a coastal crustal block similar in character to the West African Craton exists in the subsurface west of the Mauritanian anomaly. The two crustal blocks are therefore separated by the nearly continuous positive Mauritanian anomaly which likely marks a remnant, westward-dipping suture zone. Intracontinental rift sequences occur east of the suture throughout the Mauritanide and Bassaride orogens.

8.2.4 Geodynamic models

8.2.4.1 Mauritanide and Bassaride orogens　Field relationships, regional variations in radiometric ages, igneous geochemical affinities, and geophysical characteristics suggest a polyphase tectono-thermal evolution for the Mauritanide and Bassaride orogens. The following tectonic development is proposed (Figure 8.5):

(1) Middle–late Proterozoic (*c.* 1100–700 Ma; Figure 8.5A): Deposition of continental sedimentary sequences (West African Supergroup 1 of Lécorché *et al.*, 1983) on crystalline basement of the West African Shield.

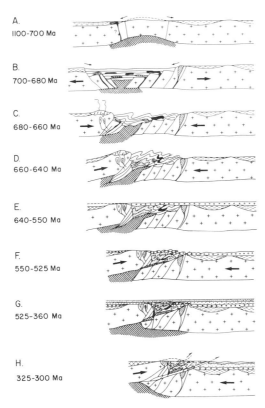

Figure 8.5　Tectono-thermal evolution of the Mauritanide and Bassaride orogens, West Africa (from Dallmeyer and Lécorché, 1989). Refer to text for discussion.

(2) Late Proterozoic (c. 700–680 Ma; Figure 8.5B): Intracontinental rifting with eventual formation of oceanic lithosphere of uncertain extent. Initial phases of extension were probably marked by emplacement of dykes of low-Ti continental tholeiite (protoliths of Farkâka amphibolite) into continental margin sedimentary sequences (Farkâka quartzites, etc.). Blanc (1986) reported 680 ± 10 Ma U-Pb zircon crystallization ages for various phases of a syenitic suite considered to have been emplaced during the late Proterozoic rifting.

Attenuation of continental crust may have resulted in partial melting and formation of alkalic basalts (Remy, 1987). These appear to have locally evolved to T-type mid-ocean ridge basalts (MORB) and formation of oceanic lithosphere (represented, in part, by fragmented ophiolitic units comprising the infrastructural Oued Amoûr and El Aoueïja units in the central Mauritanides; Table 8.1). A slightly older (720–700 Ma) compressional tectono-thermal event may be locally recorded within rift facies assemblages (Farkâka Association) in infrastructural units of the central Mauritanides (Dallmeyer and Lécorché, 1989). This may have developed while the western crustal block (Senegal microplate) was separated from the West African Craton. It probably resulted from imbrication of sedimentary sequences along the margin of the western block during initial phases of west-directed subduction.

(3) Late Proterozoic (c. 680–660 Ma; Figure 8.5C): Compression and resultant imbrication of oceanic and continental crust during Pan-African I orogenesis. This led to partial melting at deep structural levels and formation of a western, ensialic volcanic arc now represented within western infrastructural allochthons. Bassot and Caen-Vachette (1983), Dallmeyer and Villeneuve (1987) and Dallmeyer and Lécorché (1990b) reported Rb-Sr whole-rock isochron and ^{40}Ar/^{39}Ar hornblende ages suggesting c. 685–680 Ma crystallization ages for calc-alkaline granites of this suite.

(4) Late Proterozoic–early Cambrian (c. 660–550 Ma; Figures 8.5D and 8.5E): Uplift and cooling of the Pan-African I orogen. Deposition of foreland flysch sequences with basal tillite (S2a: e.g., Tichilit el Beïda and Mali Groups).

(5) Cambrian (c. 550–525 Ma: Figure 8.5F): Deformation and metamorphism in the Bassarides (Pan-African II orogenesis) with an apparent southward increase in intensity. This may record distal effects of a continent–continent collision which led to formation of the Rokelide orogen. Associated uplift and erosion resulted in deposition of foreland molasse (S2b: e.g., Youkoukoun-Taban and Mejeria Groups).

(6) Cambrian–Late Devonian (c. 525–360 Ma; Figure 8.5G): Deposition of clastic successions in the Taoudeni and Bové basins (S2b and S3).

(7) Early carboniferous (c. 325–300 Ma; Figure 8.5H): Collision of Gondwana and Laurentia results in eastward relative movement of the western continental block. This produced: (a) variable metamorphism and imbrication

of pre-deformed western sequences comprising parts of the suprastructural nappes; (b) local ductile imbrication of previously deformed and metamorphosed late Proterozoic rift sequences within infrastructural allochthons; (c) folding and thrusting of intracontinental foreland sequences (foreland fold and thrust belt and, in part, the parautochthonous external nappes of Lécorché *et al.*, 1983); and (d) emplacement of suprastructural allochthons.

(8) Early Permian (*c.* 300–275 Ma): Local ductile deformation along faults bordering the West African coastal blocks during terminal stages of its eastward translation.

8.2.4.2 Rokelide orogen   A penetrative record of Pan-African II orogenesis occurs throughout the Rokelide orogen. There are no surviving traces of potential Pan-African I metamorphism and/or deformation, and there is no evidence of late Palaeozoic orogenesis. The intense Pan-African II activity is interpreted to reflect continent–continent collision during assembly of northwestern Gondwana, and involving relative motion between the West African and Guiana Cratons. Very deep erosional levels are presently exposed in the Rokelide orogen, and, as a result, most supracrustal manifestations of oceanic closure have been removed. A suture zone may be represented by ductile shear zones (section 7, Figure 8.4) which presently separate the Kasila Group (variably retrogressed and penetratively mylonitic western gneiss terrane) and the Kenema Assemblage-Marampa Group (variably retrogressed and mylonitic structural units of West African Craton basement: e.g., Williams, 1978).

8.3   Southern Appalachian orogen

The southern Appalachian orogen may be broadly subdivided into several northeast-trending lithotectonic belts (Figure 8.6), each characterized by a distinctive group of lithologies, metamorphic grade, and/or structural style (e.g., King, 1955; Hatcher, 1972, 1978; Rankin, 1975; Glover *et al.*, 1983). Stratigraphic and/or palaeontological characteristics suggest that sequences within the Valley and Ridge and allochthonous western Blue Ridge provinces formed along the Palaeozoic margin of Laurentia. Rocks within the eastern allochthonous lithotectonic belts have uncertain palinspastic origins. Results of seismic reflection studies (e.g., Harris *et al.*, 1981) and regional gravity and magnetic characteristics (e.g., Hatcher and Zietz, 1978, 1980) indicate that autochthonous North American basement probably underlies most of the southern Appalachian orogen.

Lithostratigraphic sequences in the eastern Blue Ridge and throughout the Piedmont are not comparable with successions of similar age in either the Valley and Ridge or the western Blue Ridge which appear to have originated along the eastern margin of Laurentia. The presence of Atlantic province

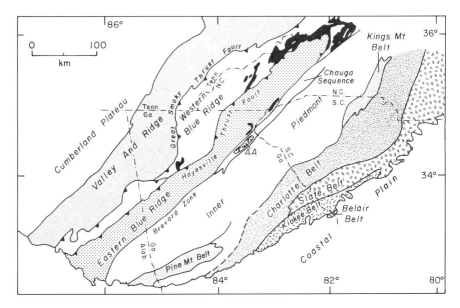

Figure 8.6 Generalized tectonostratigraphic map of the southern Appalachian orogen locating geographic and geologic features discussed in the text (adapted from Secor *et al.*, 1986a). Grenville (late Proterozoic) basement rock within the Blue Ridge are shown (black): AA = Alto allochthon.

trilobite fauna (Secor *et al.*, 1983) within the Carolina Slate belt requires that at least this (and the petrogenetically linked Charlotte belt) was faunally isolated from Laurentia during the Middle Cambrian (Secor *et al.*, 1983). Together these characteristics combine to suggest that at least the eastern Piedmont represents an exotic terrane (the Carolina Terrane of Secor *et al.*, 1983) that was accreted to North America subsequent to the Middle Cambrian. The timing of accretion is uncertain. In the Albemarle area of North Carolina, Noel *et al.* (1988) reported $^{40}Ar/^{39}Ar$ whole-rock plateau ages of *c*. 460 Ma for penetratively cleaved slate/phyllite within the Carolina Slate belt. They interpreted these to closely date cleavage formation which they suggested developed during accretion of the Carolina Slate belt to Laurentia. No clear record of this Middle Ordovician activity is obvious in any of the geochronological results presented by Dallmeyer *et al.* (1986) and Dallmeyer (1988) for the Piedmont in Georgia or South Carolina. In these areas the oldest record of tectono-thermal activity is *c*. 360–340 Ma. If this was associated with accretion of Piedmont terranes to Laurentia, distinctly different tectonic elements must comprise the Carolina Slate belt.

Three distinct late Palaeozoic deformational events affected the eastern Piedmont in Georgia and South Carolina between *c*. 315 and 270 Ma. Together these comprise the Alleghanian orogeny (Sector *et al.*, 1986a; Dallmeyer *et al.*, 1986). The first was associated with regional metamorphism of variable grade and the emplacement of felsic plutons at mid-crustal depths between $c.\,315$ and 295 Ma. A second Alleghanian event resulted in folding of early isothermal surfaces between $c.\,295$ and 285 Ma. The final phase of Alleghanian deformation led to the development of dextral ductile shear zones between $c.\,290$ and 265 Ma. Sector *et al.* (1986b) and Dallmeyer *et al.* (1986) suggest that regional variations in metamorphic grade and $^{40}Ar/^{39}Ar$ mineral cooling ages in the allochthonous eastern Piedmont of Georgia and South Carolina are a result of the exposure of variable crustal levels which developed by regional flexure during late Palaeozoic translation over thrust ramps. Their model suggests that the Charlotte and Carolina Slate belts were initially contiguous and underwent a regional metamorphism outboard of Laurentia sometime prior to $c.\,360$–340 Ma. Pre- to syn-tectonic granitic plutons were emplaced into the Carolina Slate belt between 320 and 310 Ma. This heat influx resulted in widespread metamorphism and establishment of an amphibolite grade intrastructure (Kiokee belt) and a greenschist grade superstructure. In higher crustal levels the late Palaeozoic thermal event variably reset intracrystalline argon systems which had previously cooled through closure temperatures. However, in deeper crustal levels the ambient country rock temperatures had been continuously maintained above horn-blende argon closure temperatures since initial Devonian metamorphism. This crustal section underwent westward transport onto the North American margin following emplacement of $c.\,280$–260 Ma late- to post-kinematic plutons within the intrastructure. During translation, the high grade intra-structure thickened as a result of stepping over a frontal ramp in the basal thrust. This was accompanied by dextral ductile shearing which resulted in differential offset of the high grade core and lower grade flanking zones along the polygenetic Modoc zone. Dynamically recrystallized biotite within the zone record 267 Ma $^{40}Ar/^{39}Ar$ plateau ages (Dallmeyer *et al.*, 1986) which probably date final stages of this ductile activity. Deeper crustal levels of the Charlotte belt which had been maintained above hornblende argon closure temperatures since initial Devonian metamorphism were brought to higher crustal levels as a result of regional flexure over a western, higher level thrust ramp.

The Alto allochthon within the western Piedmont appears to have been emplaced by west-directed thrusting from a Piedmont root zone following Late Devonian or earlier high grade metamorphism (Hopson, 1984; Hopson and Hatcher, 1987; Dallmeyer, 1988). This thermal chronology is consistent with work elsewhere in the western Piedmont which suggests at least local homogenization of Sr isotopes between $c.\,400$ and 380 Ma (Higgins *et al.*, 1980; Van Breemen and Dallmeyer, 1984). It is also consistent with $c.\,355$ Ma postmetamorphic $^{40}Ar/^{39}Ar$ cooling ages recorded by hornblende in western-most portions of the Inner Piedmont near Atlanta (Dallmeyer, 1978). Dallmeyer (1988) suggested that emplacement of the Alto allochthon occurred after Late Devonian or earlier high grade metamorphism and

prior to regional cooling through muscovite argon closure temperatures at
c. 315–300 Ma. Upward transport to higher crustal levels during nappe
transport is likely to be recorded by the diachronous cooling of the allochthon
through hornblende argon closure temperatures between c. 360 and 335 Ma.
Generally similar times for ductile faulting are suggested by Rb-Sr whole-
rock isochron ages reported for mylonitic rocks within thrusts bordering
several of the southern Appalachian lithotectonic belts. These include: (1) the
Brevard fault zone (356 ± 28 Ma; Odom and Fullagar, 1973); and (2) the
Great Smoky fault (368 ± 9 Ma; Hatcher and Odom, 1980).

8.4 Pre-Cretaceous crystalline basement beneath the Atlantic and Gulf Coastal Plains of the southeastern United States

The nature of the pre-Mesozoic crystalline basement beneath the Atlantic
and Gulf Coastal Plains of the southeastern United States has been revealed
by penetrations associated with deep oil test drilling. Buried extensions of
Appalachian elements (including the Valley and Ridge Province, Talledega
Slate Belt, and various Piedmont terranes) extend c. 50–60 km southeast of
the Coastal Plain unconformity (Figure 8.7). These are bordered to the south
by a series of fault-bounded basins containing Mesozoic, continental clastic
rocks which are intruded by numerous diabase dykes. Three contrasting
lithotectonic elements constitute the pre-Mesozoic crystalline basement
which has been penetrated south of the Mesozoic basins. These include
(Figure 8.7): (1) a group of metamorphic rocks of variable grade together with
deformed and retrogressed granite in southwestern Alabama and southeastern
Mississippi (Wiggins Uplift); (2) a suite of contrasting igneous rocks (granite,
basalt, and agglomerate) and serpentinite which occurs along the Brunswick-
Altamaha magnetic anomaly in southwestern Alabama; and (3) an extensive,
apparently coherent tectonic element comprised of undeformed granite,
low-grade felsic metavolcanic rocks, a suite of high grade metamorphic rocks
(gneiss and amphibolite), and a succession of undeformed, Lower Ordovician–
Middle Devonian sedimentary rocks. The latter association has been termed
the Suwannee Terrane by Thomas et al. (1989).

8.4.1 Suwannee terrane

8.4.1.1 Osceola Granite An area of undeformed granite constitutes a large
portion of the pre-Mesozoic crystalline basement of central peninsular
Florida (Figure 8.7). This has been termed the Osceola Granite by Chowns
and Williams (1983). Dallmeyer et al. (1987) describe the pluton as heterogeneous and composed dominantly of biotite granodiorite, leucocratic biotite
quartz monzonite, and biotite granite. Most of the samples examined by
Dallmeyer et al. (1987) were composed dominantly of oligoclase, quartz,

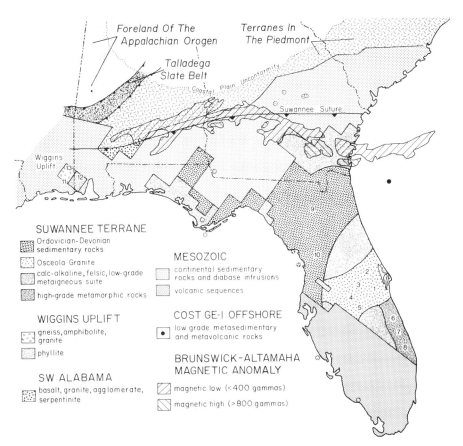

Figure 8.7   Lithotectonic units within pre-Mesozoic crystalline basement of the Atlantic and Gulf coastal plains, southeastern United States (adapted from Chowns and Williams, 1983; Thomas *et al.*, 1989). All contacts with Mesozoic sequences shown as high-angle faults. Trace of Brunswick-Altamaha magnetic anomaly from Zeitz (1982). Suwannee suture from Thomas *et al.* (1989). The wells discussed in the text are located (1–13).

perthitic alkali feldspar, and biotite. The feldspars typically display a magmatic character; plagioclase occurs as zoned crystals and alkali feldspar appears to have retained most of its original albite component in the form of perthitic lamellae. There is no textural evidence for significant subsolidus migration, exsolution, and/or recrystallization of feldspar components. Because of these petrographic characteristics, Dallmeyer *et al.* suggested the pluton experienced relatively rapid post-magmatic cooling and was probably emplaced at shallow crustal levels.

Bass (1969) reported Rb-Sr analytical results from several density fractions of feldspar from two portions of a core from a well in Osceola County (well 1, Figure 8.7). The data were extremely scattered and tentatively interpreted by

Bass to reflect a crystallization age of c. 530 Ma. Dallmeyer et al. (1987) reported $^{40}Ar/^{39}Ar$ incremental-release ages for five biotite concentrates from cuttings recovered from four wells penetrating the Osceola Granite (wells 2–5, Figure 8.7). All the samples are unaltered, and the biotite concentrates display well-defined plateau ages ranging between c. 535 and 527 Ma. Dallmeyer et al. (1987) suggested that the biotite plateau ages likely closely date emplacement of the pluton in view of its high level character and apparently rapid post-magmatic cooling.

8.4.1.2 St. Lucie Metamorphic Complex   A suite of high grade metamorphic and subordinate, associated, variably deformed igneous rocks occurs southeast of the Osceola Granite. This has been termed the St. Lucie Metamorphic Complex by Thomas et al. (1989). Predominant lithologies include amphibolite, biotite-muscovite schist and gneiss, and quartz diorite. The complex has a distinctive aeromagnetic signature (Taylor et al., 1968; Klitgord et al., 1984), with marked northwest-trending magnetic lineations. Thomas et al. (1989) suggest these may reflect structural strike.

Bass (1969) reported isotopic ages for the high grade complex, including a 503 Ma K-Ar date for a hornblende concentrate from amphibolite and a 530 Ma Rb-Sr model age for a biotite concentrate from interlayered gneiss within a well in St. Lucie County (well 6, Figure 8.7). Hornblende concentrates prepared from amphibolite cuttings recovered from wells in St. Lucie and Martin Counties (wells 7 and 8, Figure 8.7) record well-defined $^{40}Ar/^{39}Ar$ plateau ages of c. 511 and 513 Ma (Dallmeyer, 1989a). These have been interpreted to date post-metamorphic cooling through appropriate argon retention temperatures.

8.4.1.3 Felsic volcanic-plutonic complex   A felsic volcanic-plutonic complex has been penetrated in separated areas of the Coastal Plain pre-Mesozoic basement. Lithologic variants include felsic vitric tuff, felsic ashflow tuff, and tuffaceous arkose with subordinate andesite and basalt. Epizonal felsic plutons occur within some wells and are likely subvolcanic equivalents of the volcanic sequences. Mueller and Porch (1983) presented geochemical analyses for representatives of the felsic complex which suggest calc-alkaline affinities. The complex is generally undeformed, however it nearly everywhere displays low grade metamorphic assemblages.

The felsic igneous complex appears to be unconformably overlain by Lower Ordovician sandstone in one well in central peninsular Florida, and on this basis Chowns and Williams (1983) suggest a late Proterozoic–early Palaeozoic age. This is consistent with stratigraphic relationships inferred from seismic characteristics in northwestern Florida by Arden (1974). Whole-rock, K-Ar ages ranging between c. 480 and 165 Ma have been reported for many members of the felsic complex (summarized by Chowns and Williams, 1983). A representative suite of seven volcanic samples have been analyzed with whole-rock, ^{40}Ar/^{39}Ar incremental-release techniques (Dallmeyer, un-published data). All samples display markedly discordant age spectra indicating widespread disturbance of initial intracrystalline argon systems. These results suggest that the published K-Ar whole-rock ages may not be used to constrain either the time of magmatic or metamorphic events.

The COST-GE1 well was drilled *c*. 100 km east of the northernmost Florida coast (Figure 8.7) and penetrated *c* 600 m of low grade metasedimentary rocks (argillite) overlying variably metamorphosed trachyte and sandstone (Scholle, 1979). The relationship of this sequence to the mainland felsic igneous complex is uncertain. Whole-rock K-Ar ages of 374 and 346 Ma were reported for metasedimentary rocks recovered from the well (Simonis, in Scholle, 1979). A slate sample from 11 600 ft displays an internally discordant ^{40}Ar/^{39}Ar age spectrum defining a total-gas age of *c*. 341 Ma (Dallmeyer, unpublished data). A felsic metavolcanic rock from 12 350 ft also displays an internally discordant age spectrum, however intermediate- and high-temperature increments correspond to a *c*. 375 Ma plateau date. This is similar to a 363 ± 7 Ma Rb-Sr whole-rock isochron reported for seven samples from the COST well by Simonis (in Scholle, 1979). These Devonian ages are more likely related to metamorphic overprinting than to initial magmatic events.

8.4.1.4 Palaeozoic sedimentary rocks   A succession of generally undeformed sedimentary rocks occurs in several separate areas of the Coastal Plain crystalline basement. The base of the section is marked by Lower Ordovician littoral quartz sandstones (Carroll, 1963). These are overlain with presumed conformity by Ordovician to Middle Devonian shales with locally significant horizons of siltstone and sandstone. A nearly continuous succession appears to be present, however Cramer (1973) noted that the absence of Lower Silurian faunas may indicate a disconformity. Cold water, Gondwanan palaeontological affinities are displayed by all fauna throughout the entire Palaeozoic sequence (Whittington, 1953; Andress *et al.*, 1969; Goldstein *et al.*, 1969; Cramer, 1971, 1973; Whittington and Hughes, 1972; Pojeta *et al.*, 1976). Opdyke *et al.* (1987) reported *c*. 1800–1650 Ma U-Pb zircon ages for detrital zircons within a core of Ordovician–Silurian sandstone from a well penetrating the north Florida basin in Alachua County, Florida (well 9, Figure 8.7). Opdyke *et al.* presented palaeomagnetic results from the sandstone core which suggest a palaeolatitude of *c*. 49°. Dallmeyer (1987) reported a ^{40}Ar/^{39}Ar plateau age of *c*. 504 Ma for detrital muscovite from a well penetrating Lower Ordovician sandstone in Marion County, Florida (well 10, Figure 8.7).

8.4.2   *Structure*

The spacing of basement penetrations precludes reliable determination of the nature of the contacts between the various lithotectonic units which constitute the Suwannee terrane. Applin (1951), Barnett (1975), and Chowns and Williams (1983) suggested that the Palaeozoic sedimentary sequence occupies a regional synclinal structure (termed the North Florida Basin by Thomas *et al.*, 1989). Later development of horsts and grabens during Mesozoic faulting significantly affected the subcrop distribution of basement units, particularly in south Georgia, southeastern Alabama, and northwestern Florida (e.g., Smith, 1983).

8.4.3   *Relationship to Appalachian elements*

Parts of the pre-Cretaceous crystalline basement beneath the Atlantic and Gulf Coastal Plains were initially correlated with Appalachian sequences in the Valley and Ridge province (Campbell, 1939) and eastern Piedmont (Milton and Hurst, 1965). However, the undeformed character and Gond-wanan palaeontological affinities of the Suwannee succession contrast markedly with sequences of similar age in the Valley and Ridge province. In addition, the age and character of the Osceola granite and bordering high grade metamorphic sequence are unlike that of any Appalachian elements in either the Blue Ridge or Piedmont. Because of these inconsistences, most recent workers (e.g., Chowns and Williams, 1983) have suggested that the pre-Cretaceous basement units are unrelated to exposed Appalachian tectonic elements.

8.4.4   *Wiggins Uplift*

The Wiggins Uplift is an elevated block of pre-Mesozoic crystalline rocks bordered by Mesozoic faults which has been penetrated by several wells drilled in southwestern Alabama and southestern Mississippi (e.g., Cagle and Khan, 1983). Predominant rock types include phyllite, chlorite schist, meta-sandstone, amphibolite, gneiss, and variably metamorphosed and deformed granite. K-Ar whole-rock ages ranging between *c.* 300 and 275 Ma have been reported for various lithologies within the complex (Cagle and Khan, 1983). $^{40}Ar/^{39}Ar$ incremental-release ages have been determined for several units within the Wiggins Uplift (Dallmeyer, 1989b). These include: (1) a sample of deformed granite from a well in Jackson County, Mississippi (well 11, Figure 8.7) which records a markedly discordant whole-rock age spectrum corres-ponding to a total-gas date of *c.* 138 Ma; (2) phyllite from a well in Mobile County, Alabama (well 12, Figure 8.7) which records a whole-rock plateau age of *c.* 318 Ma; and (3) plateau ages of 310 and 305 Ma for hornblende and biotite concentrates from interlayered amphibolite and gneiss within a core recovered from a well in Jackson County, Mississippi (well 13, Figure 8.7).

8.4.5   *Southwestern Alabama Igneous Complex*

Several wells have penetrated various igneous rocks (including granite, basalt, and volcanic agglomerate) along the trace of the Brunswick-Altamaha magnetic anomaly in southwestern Alabama (Neathery and Thomas, 1975; Thomas *et al.*, 1989). K-Ar whole-rock ages of *c.* 335 and 267 Ma were reported by Neathery and Thomas (1975) for granite and basalt within the complex. Thomas *et al.* (1989) reported that massive serpentinite was penetrated in a well along the southern gradient of the anomaly.

8.4.6   *Regional tectonic relations*

Available basement penetrations allow demarcation of the boundary between Appalachian sequences and the Suwannee Terrane shown in Figure 8.7 (Chowns and Williams, 1983; Thomas *et al.*, 1989). This approximately coincides with the trace of the Altamaha-Brunswick magnetic anomaly in Alabama, and Nelson *et al.* (1985a, b) suggested that the anomaly everywhere marks a suture between Appalachian elements and the Suwannee Terrane. However, traced eastward across Georgia the anomaly and subsurface terrane boundary diverge (Figure 8.7). On the basis of these relationships, Chowns and Williams (1983) suggested that although the anomaly may mark the deep crustal expression of the suture, it is likely that shallower crustal levels have been thrust northward carrying the subcrop expression of the boundary over the deeper crustal interface.

The southern boundary of the Suwannee Terrane in peninsular Florida is defined by a major fault (the Jay Fault; Smith, 1983) which is likely a projection of the Bahamas fracture zone (Klitgord *et al.*, 1983). This may connect northwestward with the Pickens-Gilberton Fault System (Smith, 1983) which can be traced into the midcontinent. A Mesozoic volcanic sequence occurs south and west of the Jay Fault. This succession probably developed in response to opening of the present Atlantic Ocean (Mueller and Porch, 1983), but developed on older continental crust (Ross *et al.*, 1986). Several fault-bounded blocks of crystalline basement with characteristics similar to that of the Suwannee Terrane appear to occur in southern Florida (Thomas *et al.*, 1989). On the basis of geophysical characteristics, Klitgord and Popenoe (1984) have also suggested that several tracts of fault-bounded continental crust occur in the Gulf of Mexico west of Florida. In addition, continental crust with Pan-African age affinities was penetrated in two DSDP holes drilled in the Gulf of Mexico northeast of Yucatan (Dallmeyer, 1984).

An extensive series of northeast-trending Mesozoic grabens is developed along the boundary between the Suwannee Terrane and the various Appalachian elements. Higgins and Zietz (1983) suggested that these developed in response to initial phases of rifting of the present Atlantic Ocean.

In northwestern Florida and southeastern Alabama a northwest-trending series of Mesozoic faults intersects the grabens, producing a complex series of smaller horst and graben structures (Smith, 1983). The Wiggins Uplift appears to be localized within one of these horsts. The relationship of the Wiggins Uplift crystalline basement to that of the Suwannee Terrane is uncertain, however it has clearly been extensively overprinted by late Palaeozoic ductile strain and metamorphism. The relationship of the south-western Alabama igneous suite to either the Suwannee Terrane or basement of the Wiggins Uplift is also uncertain.

8.5   Appalachian–West African correlations

The crystalline basement rocks penetrated beneath the Atlantic and Gulf Coastal Plains were initially correlated with successions in either the Valley and Ridge or Piedmont Provinces of the Appalachians (e.g., Campbell, 1939; Milton and Hurst, 1965). However, on the basis of the Gondwanan palaeontological affinities of the Palaeozoic sedimentary succession, correla-tions with West African sequences have been suggested by most recent workers (e.g., Wilson, 1966; Rodgers, 1970). The recent collaborative field and geochronologic studies completed in the Mauritanide, Bassaride, and Rokelide orogens of West Africa have helped resolve the tectono-thermal evolution of these areas, thereby permitting direct correlation with counter-parts comprising the Suwannee Terrane beneath the Coastal Plain of the southeastern United States. These include (Figure 8.8):

(1) Correlation of the subsurface Osceola Granite and the post-tectonic Coya Granite exposed in the northern Rokelide orogen (Guinea). Both record c. 530 Ma crystallization ages (Dallmeyer et al., 1987) and display similar petrographic characteristics. Dallmeyer et al. proposed that the two plutons were initially part of a sequence of plutons emplaced along the northwestern margin of Gondwana following a c. 550 Ma Pan-African II tectono-thermal event.

(2) Correlation of the subsurface Palaeozoic sequence in the North Florida Basin with sequences of similar age in the Bové Basin (Senegal and Guinea: Chowns and Williams, 1983; Villeneuve, 1984). This is suggested by similarities in fauna and stratigraphic successions. In addition, the c. 505 Ma $^{40}Ar/^{39}Ar$ plateau age recorded by detrital muscovite within Ordovician sandstone in the Florida subsurface suggests a metamorphic source similar in age to that within the Bassaride and Rokelide orogens. The c. 1800–1650 Ma U-Pb ages reported by Opdyke et al. (1987) for detrital zircons within Ordovician–Silurian sandstone in the Florida subsurface suggest derivation from a source similar in age to the basement of the West African Craton (e.g., Bessoles, 1977). Palaeomagnetic results from the sandstone core suggest a palaeo-latitude of c. 49° which is in marked contrast to the c. 28° palaeolatitude

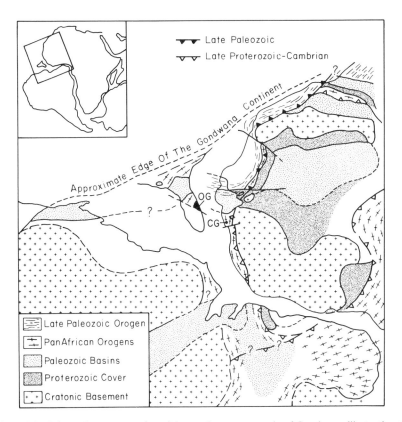

Figure 8.8 Schematic reconstruction of the northwestern margin of Gondwana illustrating the relationship of pre-Cretaceous lithotectonic units in the subsurface of the southeastern United States to correlative sequences in west Africa and northeastern South America (from Dallmeyer *et al.*, 1987). Proposed relationship of the Osceola Granite (OC) and the Coyah Granite (CG) is indicated.

suggested for Laurentia in the Ordovician–Silurian, and clearly supports a Gondwana linkage.

(3) Correlation of the subsurface St. Lucie Metamorphic Complex and portions of the Rokelide Orogen (Chowns and Williams, 1983). The effects of a penetrative *c.* 550 Ma Pan-African II tectono-thermal event are recorded throughout the Rokelide orogen. Here components of a western exotic gneissic succession are imbricated with cover sequences and penetratively mylonitic and retrogressed structural units of West African Shield basement (Allen, 1967, 1969; Thorman, 1976; Williams, 1978). In northern parts of the orogen, hornblende records K-Ar and $^{40}Ar/^{39}Ar$ postmetamorphic cooling ages ranging between *c.* 580 and 550 Ma (Allen *et al.*, 1967; Beckinsale *et al.*, 1981; Dallmeyer, 1989a). Postmetamorphic cooling appears to have been younger in the southern Rokelides where hornblende records K-Ar dates

between $c.$ 530 and 485 Ma (Hurley *et al.,* 1971; Hedge *et al.,* 1975). The $c.$ 515–510 Ma $^{40}Ar/^{39}Ar$ plateau ages recorded by hornblende within the subsurface St. Lucie Metamorphic Complex clearly support a linkage with central portions of the Rokelide orogen.

(4) Correlation of the subsurface felsic igneous complex with a calc-alkaline, variably deformed and metamorphosed igneous sequence that occurs along western portions of the Mauritanide, Bassaride, and northernmost Rokelide orogens (Dallmeyer and Villeneuve, 1987; Dallmeyer *et al.,* 1987). This sequence includes felsic volcaniclastic units together with associated, hypabyssal subvolcanic plutons. Radiometric ages suggest that the calc-alkaline igneous sequence developed between $c.$ 700 and 650 Ma (Lille, 1967; Bassot and Caen-Vachette, 1983; Dallmeyer and Villeneuve, 1987).

8.6   Terrane accretion in the southern Appalachian orogen

Geophysical, lithologic and/or faunal characteristics suggest that all southern Appalachian lithotectonic units east of the Hayesville thrust fault (Figure 8.6) are allochthonous, non-Laurentian terranes structurally overlying auto-chthonous or parautochthonous North American successions (e.g., Cook *et al.,* 1979; Harris *et al.,* 1981; Secor *et al.,* 1983, 1986b; Horton and Drake, 1986). Palaeomagnetic characteristics of plutons intruding the Inner Piedmont and the Kings Mountain, Charlotte and Carolina slate belts are not consistent with significant post-Devonian latitudinal movement. This requires that the various contrasting terranes were at least proximal to North America before the Carboniferous (e.g., Ellwood, 1982; Barton and Brown, 1983; Dooley, 1983). Sedimentological expression of outboard (eastern) tectonic instability is clearly documented in Silurian and Devonian successions deposited within the Laurentian miogeocline (e.g., Tull, 1982), and it is likely that at least some of the metamorphism and ductile thrusting recorded within the crystalline southern Appalachians is a result of initial terrane accretion to North America (e.g., metamorphism and transport of the Alto allochthon; pre-Carboniferous metamorphism and folding of the Carolina and Charlotte slate belts). Geochronological results within the same southern Appalachian belt suggest that contrasting tectonic elements may have accreted to Laurentia at different times between the Ordovician and the Devonian.

Westward transport of previously accreted terranes into their present structural positions on the North American margin occurred during the Alleghanian orogeny (Figure 8.9) which resulted from the collision of Laurentia and Gondwana (e.g., Secor *et al.,* 1986b; Dallmeyer, 1986a, b, 1988). Initial phases of Alleghanian tectono-thermal activity occurred between $c.$ 315 and 295 Ma, and involved folding, metamorphism, and emplacement of felsic plutons at middle crustal levels. The second episode of Alleghanian

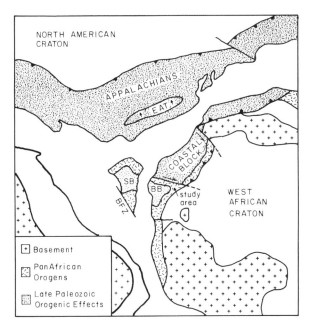

Figure 8.9   Schematic reconstruction illustrating the proposed tectonic role of the west African coastal structural block during late Palaeozoic amalgamation of Pangea: EAT = exotic Appalachian terranes which had earlier (Ordovician–Devonain) accreted to Laurentia and underwent thrust transport onto the North American margin during Late Carboniferous collision of Gondwana and Laurentia, SB = Suwannee Basin (Florida Subsurface), BB = Bové Basin, BFZ = Bahamas Fracture Zone (Mesozoic feature).

activity was associated with crustal uplift and resultant rapid postmetamorphic cooling between c. 295 and 285 Ma. This was accompanied by westward-vergent folding as crystalline nappes moved over ramps during thrust transport. Regional postmetamorphic cooling appears to have occurred slightly earlier (between c. 335 and 305 Ma) in the eastern Blue Ridge and western Piedmont. These sequences were likely maintained at elevated temperatures following a Late Devonian or earlier metamorphism which probably accompanied their initial accretion to Laurentia. Final cooling is interpreted to have occurred during transport to higher crustal levels as they were thrust onto the North American margin. The final phase of Alleghanian deformation resulted in development of dextral shear zones in the eastern Piedmont between c. 290 and 268 Ma. This strain has been interpreted to have developed as a result of relative rotation between Gondwana and Laurentia during final stages of Pangea amalgamation (Secor et al., 1986b). Eastward transport of the west African coastal block also occurred at this time, and was accompanied by development of ductile strain zones along its borders (Dallmeyer and Villeneuve, 1987). Recent continental reconstructions

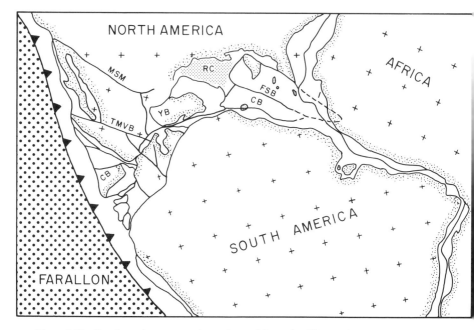

Figure 8.10   Continental reconstruction prior to Mesozoic rifting and opening of the Gulf of Mexico and Atlantic Ocean (c. 180 Ma; from Ross *et al.*, 1986). Major tectonic elements: FSB = Florida Straits block; CB (below FSB) = Cuba block; YB = Yucatan block; CB (at left) = Chortis block; MSM = Mojave-Sonora megashear; TMVB = Trans-Mexican volcanic belt. Continental crust attenuated during Mesozoic opening of Gulf of Mexico has been approximately restored to original dimensions (RC).

(e.g., Ross *et al.*, 1986; Rowley *et al.*, 1986; Ross and Scotese, 1989) suggest that final amalgamation of Laurentia and Gondwana resulted in a Pangea configuration similar to that portrayed in Figure 8.10. Fragments of Gondwana continental crust were stranded during the Mesozoic opening of the Gulf of Mexico and the Atlantic ocean (basement in the southeastern Gulf and beneath the Atlantic and Gulf Coastal Plains).

Acknowledgements

Various phases of the work summarized here were supported by grants from the U.S. National Science Foundation (EAR-8020469; EAR-8514013) and the Petroleum Research Foundation of the American Chemical Society (PRF 13920-AC2). This represents a contribution to The International Geological Correlation Program, Project 233 'Terranes In The Circum-Atlantic Palaeozoic Orogens'.

References

Bass, M. N. (1969) Petrography and ages of crystalline of basement rocks of Florida—some extrapolations. *Am. Assoc. Pet. Geol., Mem.* **11**, 283–310.
Bassot, J. P. (1966) Etude géologique du Sénégal oriental et de ses confins guinéomaliens. *Mém. Bur. Rech. Géol. Min., Paris* **40**, 322 pp.

Bassot, J. P. and Caen-Vachette, M. (1983) Donnees nouvelles sur l'âge du massif de granitoîde du Niokolo-Koba (Sénégal oriental); implication sur l'âge du stade précoce do la chaîne des Mauritanides: J. Afr. Earth Sci. 1, 159–165.

Bassot, J. P., Bonhomme, M., Roques, M. and Vachette, M. (1963) Mesures d'âges absolus sur les séries précambriennes et paléozoiques du Sénégal oriental. Bull. Soc. Geol. Fr., Ser. 7, 5, 401–405.

Beckinsale, R. D., Pankhurst, R. J. and Snelling, N. J. (1981) The geochronology of Sierra Leone (Appendix). Overseas Mem. Inst. Geol. Soc. London 7.

Bessoles, B. (1977) Géologie de l'Afrique, I. Le craton Ouest africain. Mém. Bur. Rech. Geol. Min., Paris 88, 402 pp.

Blanc, A. (1986) Le magmatism du complexe alcalin minèalisé en terres rares, Yttrium et Thorium du Bou-Naga (Mauritanie): environment géologique et géochimque, radiochronologie et significantion géodynamique. These du Doctorat, Université de Nice, 247 pp.

Bronner, G., Roussel, J., Trompette, R. and Clauer, N. (1980) Genesis and geodynamic evolution of the Taoudeni cratonic basin (Upper Precambrian and Paleozoic), western Africa. In Dynamics of Plate Interiors, Geodyn. Ser. 1, 81–90.

Bronner, G., Marchand, J. and Sougy, J. (1983) Structure en synclinal de nappes des Mauritanides septentrionales (Maroc, provinces du Sud). In Le Maroc et l'Orogenèse Paléozoique, I.G.C.P. Project No 27, Ann. Meet. Rabat, Morroco, abstract, p. 27.

Cagle, J. W. and Khan, M. A. (1983) Smackover-Norphlet stratigraphy south Wiggins Arch, Mississippi and Alabama. Trans. Gulf Coast Assoc. Geol. Soc. 33, 23–29.

Campbell, R. B. (1939) Paleozoic under Florida? Am. Assoc. Pet. Geol. Bull. 23, 1712–1713.

Carroll, D. (1963) Petrography of some sandstones and shales of Paleozoic age from borings in Florida. U. S. Geol. Surv. Prof. Pap. 454A, 15 pp.

Chiron, J. C. (1973) Etude géologique de la chaine des Mauritanide entre le parallèle de Moudjéria et le fleuve Sénégal (Mauritanie): France. Mém. Bur. Rech. Géol. Min., 84, 282 pp.

Chowns, T. M. and Williams, C. T. (1983) Pre-Cretaceous rocks beneath the Georgia Coastal Plain–regional implications. In Studies Related to the Charleston, South Carolina, Earthquake of 1886 — Tectonics and Seismicity. ed. Gohn, G. S., U.S. Geol. Surv. Prof. Pap. 1313, L1–L41.

Clauer, N. (1976) Géochimie isotopique du strontium des milleux sédimentaires: Application à la géochronologie de la couverture du craton ouest-africain. Sci. Géol. Strasbourg, 45, 256 pp.

Clauer, N., Caby, R., Jeannette, D. and Trompette, R. (1982) Geochronology of sedimentary and metasedimentary Precambrian rocks of the west African craton. Precambrian Res. 18 (1–2), 53–71.

Cramer, F. H. (1971) Position of the north Florida lower Paleozoic block in Silurian time— Phytoplankton evidence. J. Geophys. Res. 76, 4754–4757.

Cramer, F. H. (1973) Middle and upper Silurian chitinozoan succession in Florida subsurface. J. Paleontol. 47, 279–288.

Culver, S. J., Pojeta, J. and Repetski, J. E. (1988) First record of Early Cambrian shelly microfossils in West Africa. Science.

Dallmeyer, R. D. (1978) ^{40}Ar/^{39}Ar incremental-release ages of hornblende and biotite across the Georgia inner Piedmont: their bearing on late Paleozoic–early Mesozoic tectono-thermal history. Am. J. Sci. 278, 124–149.

Dallmeyer, R. D. (1984) ^{40}Ar/^{39}Ar ages from a pre-Mesozoic crystalline basement penetrated at Holes 537 and 538A of the Deep Sea Drilling Project Leg 77, southeastern Gulf of Mexico: Tectonic implications. In Initial Reports of the Deep Sea Drilling Project, Vol. 77, eds. R. T. Buffler and W. Schlager, Washington, D.C., 496–504.

Dallmeyer, R. D. (1986a) Contrasting accreted terranes in the Southern Appalachians and Gulf Coast Subsurface. Geol. Soc. Am., Abstr. Programs 18 (6), 578.

Dallmeyer, R. D. (1986b) Polyphase terrane accretion in the southern Appalachians. Geol. Soc. Am. Abstr. Programs 18 (6), 579.

Dallmeyer, R. D. (1987) ^{40}Ar/^{39}Ar age of detrital muscovite within Lower Ordovician sandstone in the Coastal Plain basement of Florida: Implications for West African terrane linkages. Geology 15, 998–1001.

Dallmeyer, R. D. (1988) Late Palaeozoic tectono-thermal evolution of the western Piedmont and eastern Blue Ridge, Georgia: Controls on the chronology of terrane accretion and transport in the southern Appalachian orogen. Bull. Geol. Soc. Am. 100, 702–713.

Dallmeyer, R. D. (1989a) A tectonic linkage between the Rokelide orogen (Sierra Leone) and the St. Lucie metamorphic complex in the Florida subsurface. *J. Geol.* **89**, 183–195.

Dallmeyer, R. D. (1989b) $^{40}Ar/^{39}Ar$ ages from subsurface crystalline basement of the Wiggins uplift and southwesternmost Appalachian Piedmont: Implications for late Paleozoic terrane accretion during assembly of Pangea. *Am. J. Sci.* **289**, 812–838.

Dallmeyer, R. D. and Lécorché, J. P. (1989) $^{40}Ar/^{39}Ar$ mineral age record of polyphase tectono-thermal evolution within the central Mauritanide orogen, West Africa. *Geol. Soc. Am. Bull.* **101**. 55–70.

Dallmeyer, R. D. and Lécorché, J. P. (1990a) Polyphase tectono-thermal evolution of the northern Mauritanide orogen (Akjoujt region): Evidence from $^{40}Ar/^{39}Ar$ Mineral Ages. *Tectonophysics* (in press).

Dallmeyer, R. D. and Lécorché, J. P. (1990b) $^{40}Ar/^{39}Ar$ polyorogenic mineral age record within the southern Mauritanide orogen (M'Bout–Bakel Region), West Africa. *Am. J. Sci.* (in press).

Dallmeyer, R. D. and Villeneuve, M. (1987) $^{40}Ar/^{39}Ar$ mineral age record of polyphase tectono-thermal evolution in the southern Mauritanide orogen, southeastern Senegal. *Geol. Soc. Am. Bull.* **98**, 602–611.

Dallmeyer, R. D., Wright, J. E., Secor, D. T. and Snoke, A. W. (1986) Character of the Alleghanian orogeny in the southern Appalachians: Part II. Geochronological constraints on the tectonothermal evolution of the eastern Piedmont in South Carolina. *Bull. Geol. Soc. Am.* **97**, 1329–1344.

Dallmeyer, R. D., Caen-Vachette, M. and Villeneuve, M. (1987) Emplacement age of post-tectonic granites in southern Guinea (West Africa) and the peninsular Florida subsurface: Implications for origins of southern Appalachian exotic terranes. *Geol. Soc. Am. Bull.* **99**, 87–93.

Deynoux, M. (1978) Les formation glaciaires du Précambrien terminal et de la fin de l'Ordovicien en Afrique de l'Ouest. Thèse de Doctorat d'Etat, Université Aix-Marseille III, 554 pp.

Dia, O. (1984) La chaîne Panafricaine et Hercynienne des Mauritanides face au bassin proterozoique supérieur à dévonien de Taoudeni dans le secteur cléf de Májéria (Taganet, sud RIM): Lithostratigraphie et tectonique. Un exemple de tectonique tangentielle superposee. Thèse Doctorat d'Etat, Marseille, 516 pp.

Dia, O., Sougy, J. and Trompette, R. (1969) Discordance de ravinement et discordance angulaire dans le Cambro-Ordovicien de la région de Méjéria (Taganet occidental, Mauritanie). *Bull. Soc. Géol. France* **7**, 267–271.

Dia, O., Lécorché, J. P. and Le Page, A. (1979) Trois événements orogéniques dans les Mauritanides d'Afrique occidentale. *Rev. Géogr. Phys. Géol. Dyn. Paris* **21** (5), 403–309.

Dooley, R. E. (1983) Paleomagnetism of some mafic intrusions in the South Carolina Piedmont. I. Magnetic systems with single characteristic directions. *Phys. Earth Planet. Inter.* **31**, 241–268.

Dupont, P. L., Villeneuve, M. and Lapierre, H. (1984) Mise en évidence de reliques océaniques au sein de la chaîne panafricaine des Mauritanides dans la région des Bassaris (Guinéé-Sénégal). *C. R. Acad. Sci. Paris* **299** (2)2, 65–70.

Ellwood, B. B. (1982) Paleomagnetic evidence for the continuity and independent movement of a distinct major crustal block in the southern Appalachians. *J. Geophys. Res.* **87**, 5339–5350.

Giraudon, R. and Sougy, J. (1963) Position anormale du socle granitise des Hagar Dekhem sur la serie d'Akjoujt et participation de ce socle a l'edification des mauritanides hercyniennes (Mauritanie occidentale). *C. R. Acad. Sci. Paris* **257**, 937–940.

Glover, L., III, Speer, J. A., Russell, G. S. and Farrar, S. S. (1983) Ages of regional metamorphism and ductile deformation in the central and southern Appalachians. *Lithos*, **16**, 223–245.

Goldstein, R. F., Cramer, F. H. and Andress, N. E. (1969) Silurian chitinozoans from Florida well samples. *Gulf Coast Assoc. Geol. Soc., Trans.* **19**, 377–384.

Harris, L. D., Harris, A. G. and de Witt, W., Jr. (1981) Evaluation of the southeastern overthrust belt beneath the Blue Ridge-Piedmont thrust. *Am. Assoc. Pet. Geol. Bull.* **65** (12), 9–36.

Hatcher, R. D., Jr. (1972) Developmental model for the southern Appalachians. *Geol. Soc. Am. Bull.* **83**, 2735–2760.

Hatcher, R. D., Jr. (1978) The Alto allochthon: A major tectonic feature of the Piedmont of northeast Georgia. *Ga Geol. Surv., Bull.* **93**, 83–86.

Hatcher, R. D., Jr. and Odom, A. L. (1980) Timing of thrusting in the southern Appalachians, U.S.A.: Model for orogeny? *Q. J. Geol. Soc. London* **137**, 321–327.

Hatcher, R. D., Jr. and Zietz, I. (1978) Thin crystalline thrust sheets in the Southern Appalachian

Inner Piedmont and Blue Ridge: Interpretation based upon regional aeromagnetic data. *Geol. Soc. Am., Abstr. Programs*, **70**, 417.

Hatcher, R. D., Jr. and Zietz, I. (1980) Tectonic implications of regional aeromagnetic and gravity data from the southern Appalachians. In *Proceedings, The Caledonides in the U.S.A.*, ed. Wones, D. R., Blacksburg, Department of Geological Sciences, Virginia Polytectonic Institute and State University, Mem. 2, 235–244.

Hedge, C. E., Marvin, R. F. and Naser, C. W. (1975) Age provinces in the basement rocks of Liberia. *J. Res., U.S. Geol. Surv.* **3**, 425–429.

Higgins, M. W. and Zietz, I. (1983) Geologic interpretation of geophysical maps of the pre-Cretaceous 'basement' beneath the Coastal Plain of the southeastern United States. *Geol. Soc. Am. Mem.* **158**, 125–130.

Higgins, M. W., Atkins, R. L. and Dooley, R. E. (1980) Structure and stratigraphy of the Atlanta area, Georgia. *Geol. Soc. Am., Abstr. Programs* **12**, 180.

Hopson, J. L. (1984) Stratigraphy and structure of the Alto allochthon, Ayersville Quadrangle, Georgia. Unpublished Ms. Thesis, University of South Carolina, 151 pp.

Hopson, J. L. and Hatcher, R. D., Jr. (1988) Tectono-thermal evolution of the Alto Allochthon of the Southern Appalachian orogen. *Bull. Geol. Soc. Am.* **100**, 339–350.

Horton, J. W. and Drake, A. A. (1986) Tectonostatigraphic terranes and their boundaries in the central and southern Appalachians. *Geol. Soc. Am. Abstr. Programs* **18** (6), 636.

Hurley, P. M., Leo, G. W., White, R. W. and Fairbairn, H. W. (1971) Liberian Age province (about 2700 Ma) and adjacent provinces in Liberia and Sierra Leone. *Geol. Soc. Am. Bull.* **82**, 1004–1005.

Kessler, S. (1986) Etude structurale et pétrologique sur les nappes internes des Mauritanides dans la région d'Akjoujt (R.I. de < airotamoe). Thèse Université d'Aix-Marseille III, **108**, 91 pp.

King, P. B. (1955) A geologic section across the southern Appalachians: An outline of the geology in the segment in Tennessee, North Carolina and South Carolina. In *Guides to Southeastern Geology*, ed. Russell, R. J., *Geol. Soc. Am.* 332–373.

Klitgord, K. D. and Popenoe, P. (1984) Florida: A Jurassic transform plate boundary. *J. Geophys. Res.* **89**, 7753–7772.

Klitgord, K. D. and Schouten, H. (1981) Mesozoic evolution of the Atlantic Caribbean and Gulf of Mexico. In *The Origin of the Gulf of Mexico and the Early Opening of the Central North Atlantic Ocean*, ed. Pilger, R. H., Proc. Symp. Houston Geol. Soc. Feb., 1981 100–101.

Klitgord, K. D., Dillon, W. P. and Popenoe, P. (1983) Mesozoic tectonics of the southeastern United States Coastal Plain and continental margin. In *Studies Related to the Charleston, South Carolina, Earthquake of 1886 — Tectonics and Seismicity*, ed. Gohn, G. S., *U. S. Geol. Surv. Prof. Pap.* **1313**, P1–P15.

Lécorché, J. P. (1980) Les Mauritanides face an craton ouest-african. Structure d'un secteur cléf: la région d'Ijibiten (Est Akjoujt, R. I. de mauritanie). Thesè Doct. Etat Univ. Aix-Mareeille III, 446 pp.

Lécorché, J. P. and Clauer, N. (1983) First radiometric data (K/Ar) on the front of the Mauritanides in the Akjoujt region (Mauritania). In *Le Maroc et l'Orogenie Paleozoique*, IGCP Project No 27, Annu. Meet. Rabat, Morocco, Abstract, p. 23.

Lécorché, J. P., Roussel, J., Sougy, J. and Guetat, Z. (1983) An interpretation of the geology of the Mauritanides orogenic belt (West Africa) in the light of geophysical data. In *Contributions to the Tectonics and Geophysics of Mountain Chains*, eds. Hatcher, R. D., Williams, H. and Zietz, I., *Geol. Soc. Am. Mem.* **158**, 131–147.

Lécorché, J. P., Dallmeyer, R. D. and Villeneuve, M. (1989) Definition of tectonostratigraphic terranes in the Mauritanide, Bassaride, and Rokelide Orogens: West Africa. In *Terranes in the Circum-Atlantic Paleozoic Orogens*, ed. Dallmeyer, R. D., *Geol. Soc. Am. Spec. Pap.* **230**, 131–144.

Lécorché, J. P., Bronner, G., Dallmeyer, R. D., Rocci, G. and Roussel, J. (1990) The Mauritanide Orogen and its northern extensions (Western Sahara and Zemmour), West Africa. In *The West African Orogens and Circum-Atlantic Correlatives*, eds. Dallmeyer, R. D. and Lécorché, J. P., Springer, Heidelberg, (in press).

Lefort, J. P. (1980) Un 'fit' structural de l'Atlantique Nord: arguments géologiques pour corréler les marqueurs geophysiques reconnus sur les marges. *Mar. Geol.* **37**, 355–369.

Legrand, P. (1969) Description de *Westonia chudeani* nov. sp., Brachiopode inarticulé de l'Adrar mauritanien (Sahara occidental). *Bull. Soc. Géol. France* **7** (11), 251–256.

Le Page, A. (1983) Les grandes unites de Mauritanides aux confins du Senegal et de la

Mauritanie: L'evolution structural de la chaine du Precambrian superieur au Devonien. These de Doctorat d'Etat, Universite d'Aix-Marseille III, 518 pp.

Lille, R. (1967) Etude géologique du Guidimakha (Mauritanie). Essai de résolution structurale d'une série épimetamorphique. *Mém. Bur. Rech. Géol. Min. Paris* **55**, 397 pp.

Marcelin, J. (1975) La chaîne des Mauritanides. Partie nord (région entre l'Akchar et l'Aouker). In *Notice Explicative de la Carte Géologique au 1/1,000,000 de la Mauritanie*, Monographies Géologiques Régionales, *Mém. Bur. Rech. Géol. Min., Paris* 117–120.

Milton, C. and Hurst, V. J. (1965) Subsurface 'basement' rocks of Georgia. *Ga Geol. Surv. Bull.* **76**, 56 pp.

Mueller, P. A. and Porch, J. W. (1983) Tectonic implications of Paleozoic and Mesozoic igneous rocks in the subsurface of Peninsular Florida. *Trans. Gulf Coast Assoc. Geol. Soc.* **33**, 169–173.

Neathery, T. L. and Thomas, W. A. (1975) Pre-Mesozoic basement rocks of the Alabama Coastal Plain. *Gulf Coast Assoc. Geol. Soc. Trans.* **25**, 86–99.

Nelson, K. D., Arnow, J. A., McBride, J. H., Willemin, J. H., Huang, J., Zheng, L., Oliver, J. E., Brown, L. D. and Kaufman, S. (1985a) New COCORP profiling in the southeastern United States. Part I. Late Paleozoic suture and Mesozoic rift system. *Geology* **13**, 714–718.

Nelson, K. D., McBride, J. H., Arnow, J. A., Oliver, J. E., Brown, L. D. and Kaufman, S. (1985b) New COCORP profiling in the southeastern United States. Part II. Brunswick and east coast magnetic anomalies, opening of the north-central Atlantic Ocean. *Geology* **13**, 718–721.

Noel, J. R., Spariosu, D. J. and Dallmeyer, R. D. (1988) Paleomagnetism and ^{40}Ar/^{39}Ar ages from the Carolina Slate Belt, Albemarle, N.C.: Implications for terrane history. *Geology* **16**, 64–68.

Odom, A. L. and Fullagar, P. D. (1973) Geochronologic and tectonic relationships between the Inner Piedmont, Brevard zone, and Blue Ridge belts, North Carolina. *Am. J. Sci.* **273-A**, 133–149.

Opdyke, N. D., Jones, D. S., MacFadden, B. J., Smith, D. L., Mueller, P. A. and Shuster, R. D. (1987) Florida as an exotic terrane: Paleomagnetic and geochronologic investigation of lower Paleozoic rocks from the subsurface of Florida. *Geology* **15**, 900–903.

Pilger, R. H. (1981) The opening of the Gulf of Mexico: Implications for the tectonic evolution of the northern Gulf Coast. *Trans. Assoc. Geol. Soc.* **81**, 377–381.

Pojeta, J., Jr., Kriz, J. and Berdan, J. M. (1976) Silurian–Devonian pelecypods and Paleozoic stratigraphy of subsurface rocks in Florida and Georgia and related Silurian pelecypods from Bolivia and Turkey. *U.S. Geol. Surv. Prof. Pap.* **879**, 32 pp.

Ponsard, J. F. (1984) La marge du craton Ouest africain du Sénégal à la Sierra Léone: Interprétation géophysique de la chaîne panafricaine et des bassins du Protérozoique a l'Actuel. Thèse Univ. Aix-Marseille III, 198 pp.

Ponsard, J. F., Lesquer, A. and Villeneuve, M. (1982) Une suture panafricaine sur la bordure occidentale du craton ouest-africain? *C. R. Acad. Sci. Paris* **295** (2), 1161–1164.

Rankin, D. W. (1975) The continental margin of eastern North America in the southern Appalachians: The opening and closing of the Proto-Atlantic Ocean. *Am. J. Sci.* **275-A**, 298–336.

Remy, P. (1987) Le magnatisme basique des Mauritanides centrales: une ouverture océanique limitée d'âge Proterozoique supérieur en Afrique de l'Ouest. Thèse Université de Nancy, 281 pp.

Rodgers, J. (1970) *The Tectonics of the Appalachians.* Wiley Interscience, New York, 271 pp.

Ross, M. I. and Scotese, C. R. (1989) Hierarchical tectonic analysis of the Gulf of Mexico and Caribbean region. In *Tectonophysics and Geodynamics Conference Report*, ed. Scotese, C., *Spec. Vol. Am. Geophys. Union* (in press).

Ross, M. I., Scotese, C. R. and Mann, P. (1986) Computer animation of Caribbean tectono-stratigraphic development. *Geol. Soc. Am., Abstr. Programs* **18** (6), 734.

Roussel, J., Dia, O., Lécorché, J. P., Ponsard, J. F., Sougy, J. and Villeneuve, M. (1984) Panafrican to Hercynian deformations in the Mauritanides and tectonic significance of gravity anomalies. *Tectonophysics* **109**, 41–59.

Rowley, D. B., Pindell, J., Lottes, A. L. and Ziergler, A. M. (1986) Phanerozoic reconstructions of northern South America, west Africa, North America, and the Caribbean Region. *Geol. Soc. Am., Abstr. Programs* **18** (6), 735.

Scholle, P. A., ed. (1979) Geological studies of the COST GE-1 well, United States South Atlantic outer continental shelf area. *U. S. Geol. Surv. Circ.* **800**, 113 pp.

Scotese, C. R., Bambach, R. K., Barton, C., Van der Voo, R. and Ziegler, A. M. (1979) Paleozoic base maps. *J. Geol.* **87**, 217–277.

Secor, D. T., Jr., Samson, S. L., Snoke, A. W. and Palmer, A. R. (1983) Confirmation of the Carolina Slate belt as an exotic terrane. *Science* **221**, 649–651.

Secor, D. T., Jr., Snoke, A. E., Bramlett, K. W., Costello, O. P. and Kimbrell, O. P. (1986a) Character of the Alleghanian orogeny in the Southern Appalachians: Part I. Alleghanian deformation in the eastern Piedmont of South Carolina. *Geol. Soc. Am. Bull.* **97**, 1319–1328.

Secor, D. T., Jr., Snoke, A. W. and Dallmeyer, R. D. (1986b) Character of the Alleghanian orogeny in the southern Appalachians: Part III. Regional tectonic relationships. *Geol. Soc. Am. Bull.* **97**, 1345–1353.

Smith, D. L. (1983) Basement model for the panhandle of Florida. *Gulf Coast Assoc. Geol. Soc. Trans.* **23**, 203–208.

Sougy, J. (1962) Contribution à l'etude géologique des guelbs Bou Leriah (région d'Aoucert, Sahara espagnol). *Bull. Soc. Géol. France* **7** (IV), 436–445.

Taylor, P. T., Zietz, I. and Dennis, L. S. (1968) Geologic implications of aeromagnetic data for the eastern continental margin of the United States. *Geophysics* **33**, 755–780.

Thomas, W. A., Chowns, T. M., Daniels, D. L., Neathery, T. L., Glover, L. and Geason, R. J. (1989) The subsurface Appalachians beneath the Atlantic and Gulf Coastal Plains. *Geol. Soc. Am. DNAG Ser. Appalachian-Ouachita Vol.* (in press).

Thorman, C. H. (1976) Implications of klippen and a new sedimentary unit at Gibi mountains (Liberia, West Africa), in the problem of Pan African—Liberian age province boundary. *Geol. Soc. Am. Bull.* **87**, 251–268.

Tull, J. F. (1982) Stratigraphic framework of the Talladega slate belt, Alabama Appalachians. In *Tectonic Studies in the Talladega and Carolina Slate Belts, Southern Appalachian Orogen*, eds. Bearce, D. N., Black, W. W., Kish, S. A. and Tull, J. F., *Geol. Soc. Am. Spec. Pap.* **191**, 3–18.

Van Breemen, O. and Dallmeyer, R. D. (1984) The scale of Sr isotopic diffusion during postmetamorphic cooling of gneisses in the Inner Piedmont of Georgia, southern Appalachians. *Earth Planet. Sci. Lett.* **68**, 141–150.

Van der Voo, R. (1983) Paleomagnetic constraints on the assembly of the Old Red Continent. *Tectonophysics* **91**, 271–283.

Van der Voo, R., Mauk, F. J. and French, R. B. (1976) Permian–Triassic continental configurations and the origin of the Gulf of Mexico. *Geology* **4**, 177–180.

Villeneuve, M. (1984) Etude géologique sur la bordure sud-ouest du craton ouest-Africain: La suture Panfricaine et l'évolution des bassins sedimentares protérozoiques et paléozoiques de la marge nw du continent de Gondwana. *Thèse Doct. d'Etat, Univ. Aix-Marseille III*, 552 pp.

Villeneuve, M. and Dallmeyer, R. D. (1987) Geodynamic evolution of the Mauritanide, Bassaride, and Rokelide Orogens (West Africa). *Precambrian Res.* **37**, 19–28.

Whittington, H. B. (1953) A new Ordovician trilobite from Florida. *Harv. Mus. Comparative Zoology Breviora*, **17**, 6.

Whittington, H. B. and Hughes, C. P. (1972) Ordovician geography and faunal provinces deduced from trilobite distribution. *R. Soc. London, Philos. Trans. Ser. B* **263**, 235–278.

Williams, H. R. (1978) The Archaean geology of Sierra Leone. *Precambrian Res.* **6**, 251–268.

Williams, H. R. (1979) An Archaean suture in Sierra Leone. *Nature (London)* **286**, 608–609.

Wilson, J. T. (1966) Did the Atlantic close and then re-open? *Nature (London)* **211**, 676–681.

Zietz, I. (1982) Composite magnetic anomaly map of the United States. *U.S. Geol. Surv. Map GP-954A*, scale 1:2 500 000.

9 The Avalon Zone type area: southeastern Newfoundland Appalachians

S. J. O'BRIEN, D. F. STRONG and A. F. KING

9.1 Introduction

Kay and Colbert (1965) introduced the name Avalon Platform to denote the easternmost element of Williams' (1964) tripartite subdivision of the Appalachian Orogen in Newfoundland. A variety of names denoting this subdivision have since been proposed (Zone H of Williams *et al.*, 1972; Avalon Zone of Williams, 1979; Avalon Terrane of Williams and Hatcher, 1983); the usage of Williams (1979) is retained here. The Precambrian and early Palaeozoic rocks of the Avalon Zone lie outboard of the miogeocline and mobile belt of the Appalachian Orogen, and were uncoupled from the latter during most of the early Palaeozoic. The Avalon Zone in Newfoundland has a unique tectonostratigraphic character defined, in part, by: (1) a broad regional homogeneity of late Precambrian stratigraphy; (2) a widespread dominantly felsic igneous event of late Precambrian age, succeeded by extensive marine sedimentation; (3) a regional sub-Cambrian unconformity; and (4) a Cambrian to lower Ordovician platformal sequence containing Acado-Baltic trilobite fauna.

On land, the zone extends eastward from the Dover and Hermitage Bay faults for approximately 200 km, making it the largest exposed segment of late Precambrian–early Palaeozoic 'Avalonian' crust in the Appalachian–Caledonian Orogen. Superb exposure of thick, well-preserved late Precambrian stratigraphic successions, granites, sub-Cambrian unconformities and profusely fossiliferous early Palaeozoic platform rocks, along with a long history of geologic investigation, have qualified this region as the Avalonian type area. The Avalon Zone's distinctive aeromagnetic signature of broad, arcuate, sub-parallel highs and lows indicate its rocks are probably continuous southeastward, for approximately 400 km, across the eastern Canadian continental shelf (Haworth and Lefort, 1979; Lefort *et al.*, 1988). This is supported by the presence offshore of late Precambrian sedimentary rocks at the Virgin Rocks (Lilly, 1966) and late Precambrian granite at the Flemish Cap (Pelletier, 1971; King *et al.*, 1985). Offshore geophysical data indicate that the zone extends southward to the east-trending Collector magnetic anomaly (Haworth and Lefort, 1979; Zeitz *et al.*, 1980; Lefort *et al.*, 1988).

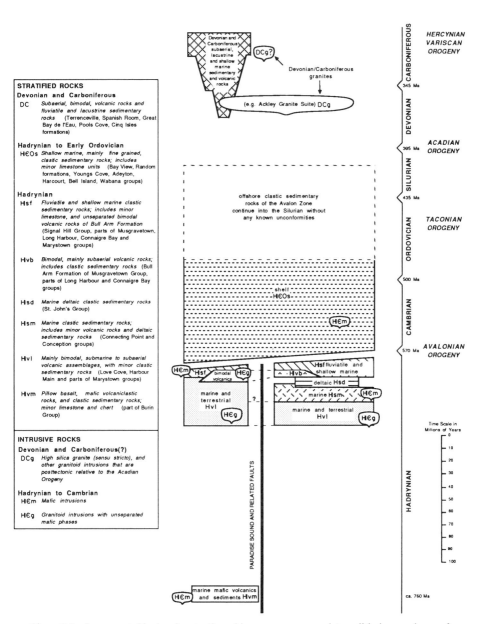

Figure 9.1 Summary table showing stratigraphic sequence, nomenclature, lithology and ages of major geological units in the Avalon Zone in Newfoundland. (Hadrynian = North American usage referring to period between 1000 Ma and Cambrian).

9.2   Characteristic lithology and age

Although the Avalon Zone in Newfoundland records more than 400 Ma of crustal evolution, its characteristic rock types formed in two relatively short time periods in the late Precambrian and Cambrian (Figure 9.1). The oldest dated rocks southeast of the Dover–Hermitage Bay fault system are part of a fault-bounded terrane consisting of c. 760 Ma submarine mafic volcanic and related clastic rocks, ultramafic slivers, tholeiitic gabbro and carbonate olistostrome (Strong et al., 1978; Krogh et al., 1988). There is no direct genetic relation between these rocks and subsequent Precambrian events, as they are separated by an apparent hiatus of 130 Ma. Between approximately 635 and 570 Ma (Krogh et al., 1988), the Avalon Zone was the site of widespread, dominantly bimodal volcanism and comagmatic plutonism, products of which are hallmarks of the zone throughout the Appalachian Orogen. Late Precambrian marine turbidites interdigitate with and overlie volcanic rocks, recording a complex history of basin evolution (Knight and O'Brien, 1988) and, locally, subsequent delta progradation (Williams and King, 1979; King, 1980). Throughout most of the region, these rocks were inhomogeneously deformed and metamorphosed under greenschist facies conditions prior to the onset of terrestrial sedimentation and associated bimodal volcanism in the latest Precambrian (Hayes, 1948; Anderson et al., 1975; O'Brien and Knight, 1988). West of the Paradise Sound Fault, the terrestrial sedimentation and volcanism, restricted largely to fault-bounded pull-apart basins, gave way first to alternating marine and non-marine sedimentation and subsequently, to deposition on a broad cratonic margin (O'Brien et al., 1977, 1984; Hiscott, 1982; Smith and Hiscott, 1984).

Across much of the eastern Avalon Zone, an erosional unconformity separates various elements of the Precambrian succession from a transgressive sequence of Lower Cambrian to Lower Ordovician platformal shale, sandstone and minor carbonate rocks (Hutchinson, 1962), which form part of the larger Acado-Baltic province (Henningsmoen, 1969).

The record in the onshore stratigraphic column for the post-Arenig–pre-Devonian interval is sparse, indicating only minor mafic magmatic activity during this period (Greenough, 1984). However, a 4000 m-thick conformable sequence of gently folded Ordovician to Silurian marine shales occurs offshore (King et al., 1986; Durling et al., 1987). In the Devonian and early Carboniferous, granitoids were intruded into, and coarse grained terrestrial clastic and alkaline volcanic rocks were deposited unconformably on, deformed older Palaeozoic and Precambrian rocks. The Devonian strata are the oldest sedimentary rocks in the Avalon Zone that contain detritus clearly linked to other terranes or zones within the orogen. The only known post-Carboniferous rocks exposed on land are Mesozoic mafic dykes (Hodych and Hayatsu, 1980). In the offshore, late Palaeozoic sedimentary basin-fills are succeeded by thick Mesozoic and younger sedimentary and locally volcanic

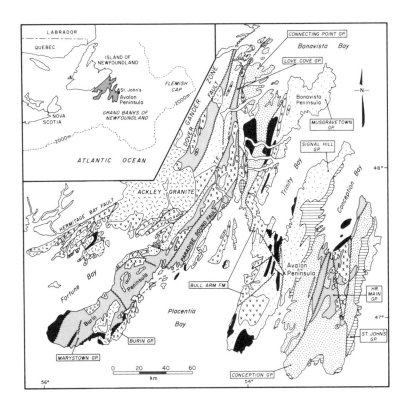

DEVONIAN and CARBONIFEROUS

⬚ High silica granite and other granitoid intrusions

⬚ Fluviatile and lacustrine clastic sedimentary rocks; terrestrial volcanic rocks (Terrenceville, Spanish Room, Great Bay de l'Eau, Pools Cove and Cinq Isles Formations)

DEVONIAN ?

⬚ Mafic intrusions

HADRYNIAN to EARLY ORDOVICIAN

⬛ Shallow marine, mainly fine grained clastic sedimentary rocks; includes minor limestone units (Bay View and Random formations; Youngs Cove, Adeyton, Harcourt, Bell Island and Wabana groups)

HADRYNIAN

⬚ Mafic intrusions

⬚ Granitoid intrusions containing unseparated mafic phases

⬚ Fluviatile and shallow marine clastic sedimentary rocks; includes minor limestone and bimodal volcanic rocks (Musgravetown Group and unseparated parts of the Long Harbour, Connaigre Bay and Long Harbour groups)

⬚ Bimodal, mainly subaerial volcanic rocks; includes clastic sedimentary rocks (Bull Arm Formation of the Musgravetown Group and parts of the Long Harbour and Connaigre Bay groups)

⬚ Marine deltaic clastic sedimentary rocks (St. John's Group)

⬚ Marine clastic sedimentary rocks; includes minor volcanic rocks (Conception and Connecting Point groups)

⬚ Terrestrial and possibly shallow marine epiclastic sedimentary rocks and associated pyroclastic rocks (Grandy's Pond litharenite)

⬚ Submarine to subaerial volcanic assemblages, with minor clastic sedimentary rocks (Harbour Main Group and parts of the Love Cove and Marystown groups)

⬚ Pillow basalt, mafic volcaniclastics, carbonate breccias and clastic sedimentary rocks; gabbro (Burin Group)

Figure 9.2   Geology of the Avalon Zone in Newfoundland, compiled (with revision) mainly from 1:250 000 maps of Jenness (1963), O'Brien *et al.* (1981), and King (1988) (TF = Terrenceville Fault).

sequences related to North Atlantic rifting and subsequent development of the continental margin (*cf.* Cutt and Laving, 1977).

9.3 Boundaries

Only the northwest boundary of the zone occurs on land in Newfoundland, and it is demarcated by the Dover Fault and the Hermitage Bay Fault (Figure 9.2). These structures lie along strike, separated by the post-tectonic Ackley Granite suite. The 368 ± 4 Ma age (^{40}Ar/^{39}Ar; biotite) from a pluton within it, which cross-cuts the fault (Kontak *et al.*, 1988), provides the minimum age of accretion of the Avalon Zone to inboard elements of the orogen.

The Dover Fault (Younce, 1970; Blackwood and Kennedy, 1975) extends from the northwest coast of Bonavista Bay, southwest for 100 km, to the Ackley Granite. It separates late Precambrian greenschist facies volcanic rocks and comagmatic granitoids from amphibolite facies gneisses, metasedimentary rocks and granites (collectively termed the Gander Zone by Williams, 1979) to the west. In its type area in and around Dover, the fault is a 500 m-wide zone of mylonite that is locally brecciated by late fault movements (Blackwood, 1977). The regional subvertical fabric of the Avalon Zone rocks adjacent to the fault is interpreted to intensify westward, culminating in mylonite in the main fault zone (Blackwood and Kennedy, 1975; Figure 9.3). The Dover Fault has a protracted history that records

Figure 9.3 Granite mylonite in the Dover Fault Zone, Dover, Bonavista Bay.

mostly strike-slip movements, the latest of which are dextral (Blackwood, 1977; Caron, 1986; Caron and Williams, 1988).

The Hermitage Bay Fault (Widmer, 1950; Blackwood and O'Driscoll, 1976) is a 50–100 m-wide cataclastic zone that extends northeast from the south coast of Newfoundland to the Ackley Granite, and separates late Precambrian volcanic, sedimentary and granitoid rocks in the southeast, from granitoids and amphibolite grade gneisses to the northwest. Rocks adjacent to the structure are overprinted by a steep fabric that regionally parallels the fault. Mylonite occurs only in a few areas (Kennedy et al., 1982).

$^{40}Ar/^{39}Ar$ release spectra (Dallmeyer et al., 1981b, 1983) on whole-rock phyllites and micas have been used to argue that the main movement on the Dover Fault, the development of fabrics in adjacent Avalon rocks and the generation of gneisses west of the fault (Gander Zone) are synchronous, and are the result of a mid-Palaeozoic, Acadian tectono-thermal event. However, recent mapping and geochronological studies in Bonavista Bay and the south coast of Newfoundland suggest that the Avalon and some of the rocks assigned to the Gander Zone may also share a pre-Acadian tectono-thermal history (O'Brien, 1987; O'Brien and Knight, 1988; Dunning and O'Brien, 1989). The major Acadian structure that marks the Avalon–Gander boundary may have developed on an older fundamental crustal weakness, possibly a major late Precambrian fault.

Deep seismic data indicate that the northwest boundary of the Avalon Zone is a vertical structure that penetrates the entire crust (Keen et al., 1986; Marillier et al., 1989). The correspondence of surface and deep crustal geology precludes the existence of a major (post-Acadian) crustal decollement under the Avalon Zone in Newfoundland and demonstrates its regionally auto-chthonous nature.

9.4   Precambrian stratigraphy

9.4.1   Pre-635 Ma rocks

Submarine mafic volcanic and sedimentary rocks and comagmatic gabbro of the Burin Group (Strong et al., 1978) are exposed in a 50 km by 5–10 km-wide, fault-bounded terrane in the southern Burin Peninsula. The group, approximately 5 km thick, is a series of pillow lava, volcanic breccia, hyalo-clasite, and fine- to coarse-grained, volcanogenic sediment (Taylor, 1976; Strong et al., 1978). A spectacular stromatolite-bearing carbonate olisto-strome unit occurs near its base. Locally, serpentinite slivers occur within fault zones that cut the group. A large comagmatic gabbro-diabase complex occurs as a stratigraphic component within the group, associated mainly with pillow lava and tuff. Basalts from the base of the succession are of alkalic chemical affinity; volcanics in the upper part of the group and the gabbro exhibit oceanic tholeiite affinity (Strong and Dostal, 1980; Figure 9.4). The

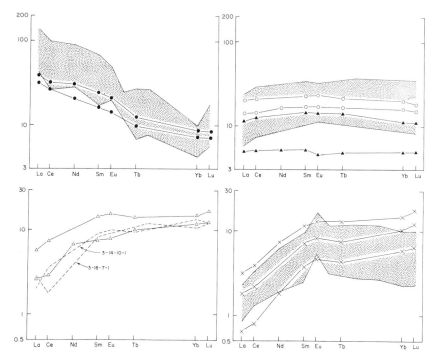

Figure 9.4   Chondrite-normalized REE patterns for the Burin Group, from Strong and Dostal (1980); top left = Pardy Island Formation, top right = Port au Bras and Path End formations, bottom left = Beaver Pond Formation, bottom right = Wandsworth gabbro.

group displays chemical characteristics and evolutionary trends that are in part similar to those of magmatic suites from modern ocean basins (Strong *et al.*, 1978; Strong and Dostal, 1980), but are also analogous, in some respects, to immature arc compositions. The Burin Group was first deformed before it was thrust against late Precambrian clastics; it is in tectonic contact with fossiliferous Cambrian strata (Strong *et al.*, 1978; O'Brien and Taylor, 1983). The comagmatic gabbro sill contains zircons that yield an age of 763 ± 2 Ma (Krogh *et al.*, 1988). The correspondence in age, composition and regional setting between the *c.* 760 Ma Pan-African ophiolitic rocks (Leblanc, 1981) and the Burin Group suggest that both formed in similar environments and represent a link between Pan-African and Avalonian belts (*cf.* Strong, 1979; O'Brien *et al.*, 1983).

9.4.2   *Late Precambrian (635 Ma–600 Ma) volcanic rocks*

Volcanic rocks that underlie thick successions of marine sedimentary rocks and that were deformed in the early pulses of the Avalonian orogeny are disposed in several regional belts across the Avalon Zone. These rocks have yielded precise zircon ages between 635 and 600 Ma (Krogh *et al.*, 1988). Their younger ages and contrasting physical and chemical characteristics indicate that there is no genetic relationship between these post-635 Ma rocks and the 763 ± 2 Ma Burin Group.

Volcanic and spatially related plutonic rocks (see below) comprise a major late Precambrian volcano-plutonic arc west of the Paradise Sound Fault (O'Brien and Taylor, 1983). These volcanic rocks form large parts of the Love Cove (Jenness, 1963), Marystown (Strong *et al.*, 1978), Connaigre Bay (Widmer, 1950) and Long Harbour groups (Williams, 1971). The Love Cove Group is a felsic to mafic, subaqueous and subaerial volcanic complex of late Precambrian age that is bounded in the east by the Paradise Sound Fault. The group consists of two belts of regionally schistose rocks, which occupy the core of two regional anticlinoria. The Marystown Group lies along strike of the Love Cove Group and comprises a volcanic complex that is, in part, lithostratigraphically equivalent. The base of neither group is exposed. Both groups are lithologically diverse suites of flows and volcaniclastic rocks that are the typical products of major caldera-forming eruptions. The lower parts of the groups contain a continuum of compositions from basalt, through andesite and rhyodacite, to rhyolite (Figure 9.5); these are of variable calc-alkaline and tholeiitic affinity (e.g., Hussey, 1979; O'Brien *et al.*, 1986). An intermediate clastic and epiclastic succession separates those rocks from a younger bimodal suite of continental, alkaline basalts and high-silica rhyolites (see below). The Love Cove Group has a U–Pb zircon age of 590 ± 30 Ma (Dallmeyer *et al.*, 1981a). The Marystown Group is disconformably overlain by uppermost Precambrian and Cambrian rocks and has been dated at

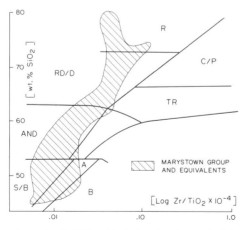

Figure 9.5   Plot of SiO_2 vs. $\log Zr/TiO_2 \times 10^{-4}$ showing the field of the Marystown Group; compositional fields after Winchester and Floyd (1977). R = rhyolites, RD/D = rhyodacites and dacites, C/P = comendites and pantellerites, TR = trachyte, AND = andesite, S/B = subalkaline basalt, A = alkali basalt, B = basanite.

608 ± 25 Ma (Dallmeyer, 1980) and $608 + 20/ - 7$ Ma (Krogh et al., 1988). The Connaigre Bay Group and its presumed lithostratigraphic equivalent, the Long Harbour Group, occur west of the Terrenceville Fault (Bradley, 1962; Figure 9.2). Both are characterized by subaerial, bimodal volcanic rocks separated by a thin, intervening sequence of shallow marine sedimentary rocks. In each group, the youngest volcanic rocks are overlain by terrestrial clastic deposits. The Connaigre Bay Group is approximately 3.3 km thick (O'Driscoll, 1977), whereas estimates of the thickness of the Long Harbour Group range as high as 10 km (Williams, 1971). Volcanic rocks of the Connaigre Bay Group and parts of the Long Harbour Group form a bimodal calc-alkaline suite (O'Driscoll and Strong, 1978). The Long Harbour Group is interpreted to form a continuous succession that passes conformably upward into fossiliferous, lowermost Cambrian strata (Williams, 1971). Neither group has been dated precisely, and the nature of their relation to each other and to volcanic successions east of the Terrenceville Fault is equivocal.

The exact relationship between the above 635–600 Ma volcanic successions and those east of the Paradise Sound Fault is unknown. On the Avalon Peninsula, volcanic rocks of late Precambrian age comprise most of what has been mapped as the Harbour Main Group (Rose, 1952). The group forms the lowest exposed part of the Precambrian succession on the Avalon Peninsula and consists of at least 1800 m of felsic to mafic, submarine and subaerial volcanic rocks. These rocks are exposed in separate fault blocks. In the eastern Avalon Peninsula, basaltic pillow lavas, pyroclastic rocks and intrusive rhyolitic domes reflect an early history of submarine volcanism and marine sedimentation (King, 1990). Flows and volcaniclastic rocks of bimodal composition, which formed mainly in a subaerial environment, are diagnostic rock types of the central Avalon Peninsula. In the group's type area at Harbour Main, subaerial ash flows, terrestrial sediments and basalt are characteristic. The latter mafic rocks are high-alumina, low-titanium basalts of transitional to mildly alkaline chemical affinities (Nixon and Papezik, 1979).

The Harbour Main Group is unconformably overlain by fossiliferous Lower Cambrian rocks (McCartney, 1967). Krogh et al. (1988) report U–Pb zircon ages of 623 ± 2 Ma and 631 ± 2 Ma from the central fault block. A 606 ± 3 Ma age from the westernmost fault block probably implies existence of stratigraphic complexities within the group, which have not yet been demonstrated in the field (Krogh et al., 1988).

9.4.3   Late Precambrian (635 Ma–c. 570 Ma) marine sedimentary rocks

The dominant feature of the Precambrian geology east of the Paradise Sound Fault is the widespread occurrence of thick sequences of marine turbidites and pelagic sedimentary rocks. The contacts of these rocks with the older volcanic successions are in most places tectonic, although conformable

(Williams and King, 1979; King, 1986, 1988; O'Brien, 1987) and, more rarely, unconformable (McCartney, 1967) contacts are documented. In the eastern Avalon Zone, deep marine turbidite sedimentation was followed by basin shoaling and delta progradation, and ultimately alluvial sedimentation. Similar-aged sedimentary rocks west of the Paradise Sound Fault are aerially restricted synvolcanic epiclastics deposited mainly in narrow, faulted basins (O'Brien *et al.*, 1984).

The Precambrian marine clastic rocks of the Avalon Peninsula form a continuous stratigraphic succession represented by the Conception Group and overlying St. John's Group (Rose, 1952; Williams and King, 1979; King, 1986). The Conception Group is a 3 km-thick sequence that consists mainly of marine siliciclastic rocks with thin tuffaceous intercalations. Near its base, the group contains distinctive submarine debris flows with a glaciogenic component, overlain by a thick succession of turbidites. These are capped by a tuffaceous sandstone–mudstone unit, which hosts a prolific and diverse late

Figure 9.6   Late Precambrian fauna: *Charniodiscus concentricus* with *Charnia masoni* Ford, Mistaken Point Formation, Conception Group, near Mistaken Point.

Precambrian (Ediacaran) fauna (Anderson and Misra, 1968; Williams and King, 1979; Anderson and Conway Morris, 1982; Figure 9.6). Gardner and Hiscott (1988) recognized three stages of deposition within the group. Early sedimentation related to westward or southwestward progradation of a submarine fan was succeeded by deposition of glaciogenic debris flows. Renewed sedimentation was related to southwestward and, subsequently, northwestward-prograding submarine fans.

The Conception Group is overlain transitionally by marine shale and thin sandstone that form the basal portion of the St. John's Group, a southward-prograding deltaic sequence. The shales also contain late Precambrian micro-fossils, including *Bavlinella* (Hofmann *et al.*, 1979).

The Connecting Point Group (Hayes, 1948) is a thick (4–5 km) sequence of marine clastic rocks that occupy a major basin in the central part of the Avalon Zone, immediately east of the Paradise Sound Fault. The Connecting Point Group conformably overlies volcanic rocks of the Love Cove Group (O'Brien and Knight, 1988) and is overlain unconformably by latest Pre-cambrian terrestrial clastic rocks (Hayes, 1948). Its relation to the Conception Group is uncertain, although both groups have generally been considered equivalent (McCartney, 1967). The group is characterized by two distinct sequences of turbidites deposited in basinal and slope settings, prograding fan lobes, and channel-levee complexes of submarine fans (Knight and O'Brien, 1988). Separation of the sediments is coincident with the introduction of mixtite into the basin and the intrusion of mafic dykes and plutons (Knight and O'Brien, 1988). The upper part of the group is rich in late Precambrian microfossils (Hofmann *et al.*, 1979).

West of the Paradise Sound Fault, comparable thicknesses of similar-aged clastic rocks are unknown. Equivalents may occur in the Anderson's Cove Formation (White, 1939) of the Long Harbour Group and equivalents in the Connaigre Bay Group and on the west side of the Burin Peninsula, but correlation is uncertain. These units are marked by sharp lateral and vertical facies variations, and define a coarsening-upward succession of shallow marine sandstones that pass upward into red clastic rocks; these may have formed in restricted fault-bounded basins (O'Brien *et al.*, 1984).

9.4.4   *Latest Precambrian terrestrial sedimentary rocks*

The Signal Hill Group conformably and disconformably overlies deltaic sedimentary rocks of the St. John's Group, but locally lies with pronounced unconformity upon deeper levels of the Precambrian clastic succession (Anderson *et al.*, 1975). It is a coarsening-upward succession of variegated siltstone and arkose and red conglomerate with a cumulative exposed thickness of 5 km (King, 1980, 1988). These sediments were deposited in an alluvial plain environment and represent the molasse to the main phases of the Avalonian orogeny (King, 1980). Lateral and vertical facies variations reflect a tectonically active basin margin to the north of the Avalon Peninsula. The Musgravetown Group, which unconformably overlies the Connecting Point Group, is a thick and lithologically variable subaerial succession (Jenness, 1963; McCartney, 1967; O'Brien and Knight, 1988). The group consists of a basal conglomerate unit, overlain by a major bimodal volcanic formation (described separately below), which is succeeded by shallow marine to terrestrial, coarsening-upward clastic sedimentary rocks (McCartney, 1967; O'Brien and Knight, 1988). In almost all areas, the group passes disconformably upward into Lower Cambrian strata; a conformable contact with possible Cambrian strata occurs in the northwesternmost Avalon Zone (O'Brien and Knight, 1988). The Musgravetown Group is stratigraphically equivalent to the Signal Hill Group in the east.

The Rencontre Formation (White, 1939) is considered to be, in part, lithostratigraphically equivalent to the other latest Precambrian clastic successions. It displays sharp vertical and lateral facies variations and is composed mainly of red siltstone, sandstone and pebble to boulder conglomerate deposited in fan delta and braided stream environments (Williams, 1971; Smith and Hiscott, 1984).

9.4.5 Latest Precambrian (c. 570 Ma) volcanic rocks

Volcanic rocks of latest Precambrian age overlie the thick marine sedimentary basins in the central Avalon Zone. They form a bimodal, lithologically variable succession of subaerial basaltic flows and pyroclastic rocks, and rhyolitic ash flow tuffs and related breccias (McCartney, 1967). The

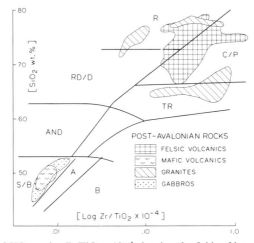

Figure 9.7 Plot of SiO_2 vs. $\log Zr/TiO_2 \times 10^{-4}$ showing the fields of late Precambrian, post-Avalonian granites and volcanics; compositional fields after Winchester and Floyd (1977) as in Figure 9.5.

volcanic rocks generally occur at or near the base of the Musgravetown Group and its equivalents in this region. The felsic rocks are a continuum from alkali rhyolite to pantellerite and comendite (Hussey, 1979; S. J. O'Brien, unpublished data; Figure 9.7), compositions that are a signature of some of the volcanic successions that post-date the Avalonian deformation recorded by the regional sub-Musgravetown unconformity.

In the northwest Avalon Zone, the peralkaline rocks are stratigraphically separable from and approximately 50 Ma younger than the volcanic suites that underlie the marine sedimentary rocks (O'Brien, 1987; O'Brien and Knight, 1988; G. R. Dunning, pers. commun., 1989). Rocks of peralkaline composition have been found in the Bull Arm Formation (Musgravetown Group) where they unconformably overlie the Connecting Point Group (Hussey, 1979; S. J. O'Brien, unpublished data).

Rocks of equivalent chemical compositions occur in the Mooring Cove Formation, near the top of the Long Harbour Group (O'Brien et al., 1984). Alkaline to mildly peralkaline compositions are also present in apparently deeper stratigraphic levels of the group (Belle Bay Formation). The overall extent of this volcanic event has yet to be demonstrated, and rocks of equivalent age probably occur elsewhere in the western Avalon zone. Much more detailed precise geochronological and geochemical data are required to fully document their distribution.

9.5   Late Precambrian plutonic rocks

The widespread occurrence of foliated, hornblende- and biotite-bearing, calc-alkaline, epizonal granite, granodiorite and diorite of late Precambrian (620–570 Ma) age is one of the most distinctive aspects of the Avalon Zone. In Newfoundland, these rocks form the core of many of the volcanic massifs (see Figure 9.2).

The Holyrood Intrusive Suite (Rose, 1952; King, 1988) of the Avalon Peninsula intrudes the Harbour Main Group and is unconformably overlain by fossiliferous Cambrian strata; the suite yields zircons dated at 620 Ma (Krogh et al., 1988). The close similarity in age between the Holyrood Intrusive Suite and the adjacent Harbour Main Group volcanics suggests they share a comagmatic relationship. On the Burin Peninsula, the granitoids form a discontinuous 120 km-long belt of elongate, northeast trending plutons, the largest of which (Swift Current Granite) has been dated (U–Pb zircon) at 580 ± 20 Ma (Dallmeyer et al., 1981a). The Hermitage Complex and Simmons Brook batholith are plutons of similar composition that intrude volcanic rocks of the westernmost Avalon Zone near the Hermitage Bay Fault (Greene and O'Driscoll, 1976; O'Driscoll and Strong, 1978). All plutons contain locally extensive dioritic and gabbroic phases, and evidence for the comingling of mafic and felsic magmas is common.

In the western Avalon Zone, stratified Precambrian rocks are also intruded by small plutons that comprise gabbro, alkali granite, riebeckite-bearing peralkaline granite and related hybrid rocks (O'Brien *et al.*, 1984; O'Brien, 1987). These plutons are spatially associated with pantellerites and comendites of late Precambrian age (O'Brien *et al.*, 1984; O'Brien, 1987) and have yielded preliminary Rb–Sr mineral and whole-rock ages between 560 and 540 Ma (J. Tuach, pers. commun., 1989). Unpublished U–Pb zircon ages (G. R. Dunning, pers. commun., 1989) together with the chemical composition of the plutons, suggest this plutonism and latest Precambrian peralkaline volcanism are synchronous.

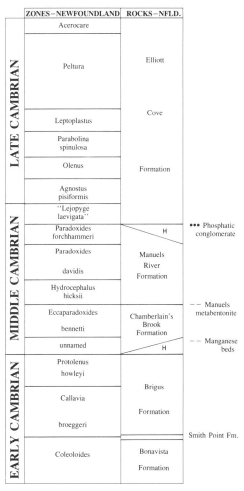

Figure 9.8 Post-Random lithostratigraphy and biostratigraphy of the Cambrian of eastern Newfoundland. Modified from Boyce (1988); zones are after Howell (1925), Hutchinson (1962), Poulson and Anderson (1975) and Martin and Dean (1981); H = hiatus.

9.6 Cambro-Ordovician stratigraphy

East of the Paradise Sound Fault, the Signal Hill and Musgravetown groups are disconformably overlain by the basal Cambrian Random Formation (Walcott, 1900; Anderson, 1981). The formation consists of quartz arenite members that, towards the west, become interbedded with an increasing amount of grey siltstone (Butler and Greene, 1976). An erosional disconformity separates the Random Formation from an overlying succession of Lower Cambrian red, green and purple shale and pink limestone that form (in ascending order) the Bonavista, Smith Point and Brigus formations of the Adeyton Group (Figure 9.8). The Brigus Formation is succeeded disconformably by the Middle Cambrian Chamberlains Brook Formation (Howell, 1925; Hutchinson, 1962), which is dominated by grey and green shale. On the northern Avalon Peninsula, black manganiferous shale occurs near its base; locally, a thin volcanic member is found in its upper part (Fletcher, 1972). Black shale of the Manuels River Formation of the Harcourt Group (Jenness, 1963) conformably overlies the Chamberlains Brook Formation. The shale, which is profusely fossiliferous (Figure 9.9), is locally interbedded with alkali basalt flows and sills (McCartney, 1967; Greenough and Papezik, 1985). A phosphatic pebble conglomerate separates the Manuels River Formation from black micaceous shale and siltstone (containing limestone concretions) of the mainly Upper Cambrian Elliot Cove Formation. In western Trinity Bay, Tremadocian grey and black shale and siltstone containing black

Figure 9.9 *Agraulos ceticephalus* Barrande from the Manuels River Formation, Manuels River (photo courtesy of W. D. Boyce; × 4 magnification).

limestone nodules and concretions (Clarenville Formation; Van Ingen, 1914) conformably overlie Upper Cambrian shale (Jenness, 1963; Martin and Dean, 1981). At Bell Island, in Conception Bay, micaceous sandstone, quartz arenite, siltstone and oolitic hematite of the Upper Cambrian(?) to Lower Ordovician Bell Island Group are succeeded by graptolitic shale, sandstone and oolitic hematite beds of the Arenigian Wabana Group (Ranger et al., 1984).

West of the Paradise Sound Fault, late Precambrian volcanic rocks are disconformably overlain by a continuous succession of nonmarine and shallow marine rocks (Widmer, 1950; Williams, 1971; O'Brien et al., 1977). The Chapel Island Formation, which conformably overlies the Upper Precambrian Recontre Formation, is a sequence of green and variegated sandstone, siltstone and mudstone, containing fossiliferous, lowest Cambrian limestone near its top (Greene and Williams, 1974; O'Brien et al., 1977). The formation is 600 m thick in its type section, and is presently under review as the stratotype for the Cambrian–Precambrian boundary (Narbonne et al., 1987). The Chapel Island Formation is conformably overlain by white quartz arenite and intercalated siltstone of the Random Formation, which in this region varies in thickness from 200 to 300 m (Butler and Greene, 1976). The Random is disconformably overlain by red and green shale and nodular limestone of the Brigus Formation; intervening Bonavista and Smith Point Formations are absent in the southwest Burin Peninsula (O'Brien et al., 1977) and elsewhere in the Fortune Bay area (Hutchinson, 1962). On the north shore of Fortune Bay, the Lower Cambrian rocks are succeeded by Middle Cambrian green and grey shale and siltstone of the Youngs Cove Formation and black shale and limestone of the Upper Cambrian Salmonier Cove Formation (Widmer, 1950).

9.7 Mid-Palaeozoic stratigraphy

The on-land mid-Palaeozoic record of the Avalon Zone is sparse and rocks of this age are known to occur only in small outliers west of the Paradise Sound Fault. The thickest successions of Devonian strata occur in the northwest Fortune Bay area (Pool's Cove, Great Bay de L'Eau and Cinq Isles formations: White, 1939; Widmer, 1950), where the sediments unconformably overlie Cambrian strata and are intruded by Devonian granites (Widmer, 1950; Williams, 1971). Elsewhere, Devonian and Carboniferous sedimentary rocks (Terrenceville and Spanish Room formations; Bradley, 1962; Strong et al., 1978) are spatially associated with major fault zones. In general, fine- to coarse grained terrestrial clastic rocks of fluviatile or lacustrine origin are dominant in all these units. These rocks contain clasts that were derived from the adjacent Gander Zone and provide a record of the earliest definitive stratigraphic linkage of the Avalon and Gander zones. Subaerial ash flow tuffs and basalt flows of Devonian age (Krogh *et al.*, 1988) overlie Precambrian rocks on the southern Burin Peninsula. These rocks (Grand Beach Complex) occur where a major northeast lineament, which strikes into the Dover Fault, intersects the Acadian flexure on the southern Burin Peninsula.

9.8 Palaeozoic plutonic rocks

Cambrian alkali gabbro occurs in widely spaced, pipe-like intrusions that occur in a narrow, north trending belt on the western Avalon Peninsula (Greenough and Papezik, 1985). Mid-Palaeozoic mafic plutonic rocks are notably rare. The only documented examples are Silurian basaltic sills that intrude the Cambrian sequence near St. Mary's Bay (Greenough, 1984).

The most widespread Palaeozoic plutonism is entirely felsic in composition and of late Devonian to earliest Carboniferous age. The plutons are typically metaluminous, high silica, biotite granites that exhibit little or no tectonic fabric (e.g., Ackley Granite; Dickson, 1983). These extensive, highly differentiated bodies cross-cut the regional fabric in both Cambrian and Precambrian rocks and locally intrude fossiliferous Devonian strata (Greene and O'Driscoll, 1976; Dallmeyer *et al.*, 1983). The granites, which locally straddle the Avalon–Gander boundary, are found mainly in a 50 km-wide corridor bounded in the west by the Dover Fault. Granites of this age do not occur on the Avalon Peninsula. The youngest known Palaeozoic intrusion in the Avalon Zone is the high-level, hypersolvus, riebeckite–aegirine alaskitic St. Lawrence granite, which has yielded an Rb–Sr isochron age of 328 ± 5 Ma (Bell and Blenkinsop, 1975).

9.9 Mesozoic magmatism

The youngest igneous rock within the Avalon Zone is a northeast trending rectilinear diabase dyke that transects the southeastern Avalon Peninsula, and is traceable into the offshore. The dyke has been dated at 201 ± 3 Ma (Hodych and Hayatsu, 1980) and has been correlated with other extensive Mesozoic intrusions in Maritime Canada (Papezik and Barr, 1981); all are related to Mesozoic rifting and the opening of the North Atlantic.

9.10   Tectono-thermal history

There is clear evidence that the Avalon Zone was affected by at least two major episodes of regional deformation and low grade metamorphism; the first is Precambrian, the second Palaeozoic. There is general agreement that the (post-635 Ma) Precambrian rocks were deformed in a regionally inhomo-

Figure 9.10 Angular unconformity of the Upper Precambrian Musgravetown Group on the Upper Precambrian Connecting Point Group, Newman Sound, Bonavista Bay.

geneous fashion and affected by low grade regional metamorphism during a late Precambrian orogenic episode known as the Avalonian orogeny (Lilly, 1966; Rodgers, 1967, 1972; Hughes, 1970). Three ages of exposed angular unconformities (sub-Conception Group, sub-Musgraveton Group, and sub-Random Formation) reflect three phases of this orogeny. The existence of yet an earlier separate orogenic phase not related to classic Avalonian events, may be recorded by the marked hiatus between the 763 ± 2 Ma Burin Group and post-635 Ma Precambrian successions.

The stratigraphically oldest Avalonian unconformity is marked by a locally erosive relationship between the Harbour Main Group and the overlying Conception Group in the Holyrood horst (McCartney, 1967). The next youngest unconformity separates latest Precambrian red beds from underlying marine clastic sedimentary rocks (Hayes, 1948; Anderson *et al.*, 1975; Figure 9.10). Folds and cleavages in the underlying rocks are truncated by the unconformity surface. Sediments above the unconformity contain metamorphic detritus, including granites (with pre-incorporation foliations) and sericite schist, as well as detrital muscovite (Jenness, 1963; O'Brien and Taylor, 1983). On the Avalon Peninsula, the uppermost part of the alluvial clastic succession is locally unconformable on the base of the older marine sedimentary rocks, and the intervening rocks including the entire deltaic sequence are missing (Anderson *et al.*, 1975). In other parts of the peninsula, the red beds are disconformable on older rocks, although there is no angular discordance (King, 1990). In the western Avalon Zone, the erosional nature of the unconformity appears to be less profound, although its angular character is clear. The youngest phase of the orogeny is marked by the unconformity at the base of the Cambrian–Ordovician platformal succession. Folded, latest Precambrian terrestrial sediments, as well as older granites, volcanic rocks and marine sediments are all unconformably overlapped by Cambrian strata. The broad, concentric distribution of the Precambrian strata around the older volcano-plutonic core of the Avalon Peninsula is, in part, the result of late Avalonian doming.

The Avalonian orogeny is generally thought to be manifest mainly by granitoid emplacement, block faulting and gentle folding, with restricted development of penetrative tectonic fabrics (McCartney, 1967; Hughes, 1970). Deformation was accompanied by low grade (prehnite–pumpellyite to chlorite) regional metamorphism. Historically, the orogeny, as recorded in its type area, was considered to be a mild disturbance. Recent work in the eastern Avalon Peninsula (King, 1986) and in the northwest Avalon Zone (O'Brien, 1987) has shown that Precambrian high strain zones and associated fabrics are more widespread than previously thought. Many of the major Acadian faults, including the Dover Fault and Paradise Sound Fault, may have been preferentially sited on earlier Avalonian high strain zones. The recently documented late Precambrian amphibolite grade rocks west of the Hermitage Bay Fault (Dunning and O'Brien, 1989; see Section 9.11) may be relics of higher grade Avalonian events preserved in a separate terrane inboard of Avalon.

Unconformities indicate that at least the western Avalon Zone was affected

Figure 9.11   Folded, Lower Cambrian Random Formation, Bar Haven, Placentia Bay.

Figure 9.12   Angular unconformity of the Devonian Great Bay de L'eau Formation on the Lower Chapel Island and Random formations and Middle Cambrian Chamberlains Brook Formation, Blue Pinion Cove, Fortune Bay (field of view is approximately 70 m).

by widespread deformation and low grade regional metamorphism between the late Cambrian and late Devonian (Figures 9.11 and 9.12). $^{40}Ar/^{39}Ar$ plateau ages between 400 Ma and 353 Ma, for whole-rock phyllites, and between 356 Ma and 352 Ma, for hornblende separates from crosscutting plutons in the western half of the zone (Dallmeyer et al., 1983), indicate that this deformation was related to Acadian, rather than Taconic, orogenic pulses. This conclusion is supported by offshore data that demonstrates the conformable transition of Ordovician and Silurian strata below the sub-Devonian unconformity (Durling et al., 1987). In general, the heterogeneous Acadian deformation resulted in widespread thrusting and strike-slip faulting and production of upright to westerly inclined folds in the western Avalon Zone, together with the development of regionally penetrative fabrics (Strong et al., 1978), locally overprinting Avalonian structures (O'Brien, 1987). The main movement along the zone-bounding Dover Fault is considered to have been Acadian (Dallmeyer et al., 1981b). In the eastern Avalon Zone, the major effects of Acadian orogenesis are open, periclinal folds and vertical faults; the structural style of this region is described in detail by King (1986, 1990).

The record of significant post-Acadian orogenic activity is scant. Devonian stratified rocks, which are of limited extent, are generally either undeformed, gently tilted or affected by open warps. However, in Fortune Bay, Devonian rocks are crosscut and bounded by faults and overthrust by Precambrian granites (Greene and O'Driscoll, 1976). The Devonian and Carboniferous plutonic rocks crosscut Acadian fabrics in both Precambrian and Cambrian rocks (Dallmeyer et al., 1981b).

9.11 Synthesis

This final section is an attempt to draw together the salient features of the Avalon Zone in Newfoundland and to discuss the broad ramifications these may have for tectonic modelling of Avalon-type terranes.

9.11.1 Deep crust

Little is known about the nature of the deep crust of the Avalon Zone, although its dominantly silicic igneous character suggests a continental component. This is supported by the general correlation of the Avalon in Newfoundland with similar-aged volcano-magmatic terranes of New Brunswick, Channel Islands, Brittany and northwest Africa, where the late Precambrian rocks display links to older gneissic basement (e.g. Strachan et al., Chapter 5 this volume). However, isotopic data from Newfoundland, which are sparse and mostly from zircons (Krogh et al., 1988; G. R. Dunning, pers. commun., 1989), do not record evidence of ancient continental crust under the Avalon Zone. A high positive εNd value obtained from a late Precambrian granite (B. J. Fryer, pers. commun., 1990; Fryer et al., 1988) intruded after the main Avalonian event, suggests a significant mantle input at this time.

Refraction data indicate that the Avalon Zone Precambrian crust is characterized by seismic velocities between 6.0 and 6.6 km/s (Sheridan and Drake, 1968). A thin, high velocity crustal layer is locally observed in the Avalon Zone, but is missing in the adjacent Gander zone; its significance is not clear. In general, the regional pattern of Bouger gravity anomalies mimics that of the magnetic anomalies, although, in detail, some on land gravity anomalies correspond to neither magnetics nor known surface geology (Miller, 1977). Deep seismic data show that the Moho beneath the Avalon Zone has a distinct topography, offset by east-dipping reflectors (Keen et al., 1986; Marillier et al., 1989). Their origin is unknown and may relate to deep crustal attenuation in the style of the Pan-African belts (cf., Kroner, 1980) or extensional tectonics in modern continental rifts (cf., Wernicke, 1985).

9.11.2 The Burin Group

These rocks are approximately 130 Ma older than any other late Pre-cambrian elements of the Avalon Zone, and constitute a separate terrane within it. The group's igneous rocks, in part, display ophiolitic chemical affinities, implying the existence of oceanic or back-arc basin crust 760 Ma ago. Its sedimentary facies may record the existence of an older, adjacent carbonate platform. The Burin Group provides a link to the Pan-African belts of Gondwana, although its unique age precludes direct genetic links with subsequent Avalonian events.

9.11.3 Volcanic and plutonic rocks

The zone's diagnostic late Precambrian volcanic and plutonic rocks formed over a very broad area in a restricted time interval. Igneous rocks that are unequivocally part of the succession below the sub-Musgravetown unconformity may have been emplaced or erupted over 20 Ma. Voluminous calc-alkaline, diorite–granodiorite–granite magmas were intruded, over a short time, into a region affected by low grade metaorphism and a relatively simple structural style. The heat source for this magmatism (and related effusive products) is a fundamental problem with important implications for Avalonian tectonics. It may be related to subduction of a mantle upwelling, perhaps a spreading ridge, as invoked by Uyeda and Miyashiro (1974) for the sea of Japan, and more recently by Wilson (1988) for other modern environments.

The earlier of the two pulses of late Precambrian volcanism is characterized by the subaerial and submarine eruptions of rhyolite, rhyodacite, basalt and, more rarely, andesite of a broad compositional spectrum. This activity was, in part, synchronous with marine sedimentation in volcanic arc basins (see below), which may have developed on either sialic or transitional crust, and is compatible with such a tectonic environment. The present chemical database alone, however, neither refutes nor confirms either of the above possibilities.

9.11.4 Sedimentary rocks

Deep marine sedimentary basins of Precambrian age are dominant features of the zone in Newfoundland and tectonic models for the Avalonian belts must accommodate their presence. They are characterized by extremely thick basin-fills, which include complex successions of marine turbidites (genetically related to volcanic rocks), deltaic facies rocks, and locally alluvial deposits. The development of marine facies in one of these basins (Eastport Basin; Knight and O'Brien, 1988) is analogous to modern marginal basin sedimentation. The late Precambrian sediments may have formed in one large basin produced by foundering of an ensialic to transitional volcano-plutonic substrate (cf., the Lord Howe Rise; Jongsma and Mutter, 1978); alternatively, they may have occupied several distinct basins, possibly separated by volcanic highs (Hussong and Uyeda, 1981). The distinct geophysical signature of the zone across the continental margin (Haworth and Lefort, 1979) may reflect an original late Precambrian configuration of linked volcanic ridges and sedimentary basins similar to the Marianas region of the Pacific Ocean (Hussong and Uyeda, 1981; Knight and O'Brien, 1988).

9.11.5   Latest Precambrian rocks

The Precambrian succession deposited above the sub-Musgravetown unconformity has at least two main characteristics. Its sedimentary rocks are terrestrial clastics deposited, at least locally, on fan-deltas and alluvial fans within fault-bounded basins that formed locally in response to strike-slip fault movements (e.g., Smith and Hiscott, 1984). The associated volcanics are terrestrial bimodal suites, and like the comagmatic plutons, are strongly alkaline or peralkaline. These differences of age, chemical composition and styles of sedimentation seen across the sub-Musgravetown unconformity, may reflect a fundamental shift towards a transtensional tectonic regime after the main pulse of the late Precambrian Avalonian orogeny, prior to the deposition of Cambrian platformal sediments. The chemical evolution is analogous to that seen in some modern continental rifts, such as the southwestern USA, where the shift to alkalic compositions is synchronous with the onset of regional transtension, which was related to the collision of the East Pacific Rise with North America (Atwater, 1970; Lipman et al., 1972). The late Precambrian evolution of the Avalon Zone is also analogous to several of the Pan-African belts of Gondwana (Strong, 1979; O'Brien et al., 1983), particularly the Afro-Arabian region (Hijaz Arc; Stern et al., 1984), where a shift from compression to transtension in the latest Precambrian has been documented. It is probable that all evolved in related or comparable tectonic environments and were assembled in a major late Precambrian tectono-thermal event.

9.11.6   The Eocambrian–Palaeozoic record

Throughout the western Avalon Zone, alluvial sedimentation in pull-apart basins in the latest Precambrian, gave way to near shore and open shelf marine deposition in a tectonically stable region (Anderson, 1981; Hiscott, 1982; Smith and Hiscott, 1984). In the early Cambrian, the quartz arenite-dominated, transgressive Random Formation was deposited, in places disconformably, upon Precambrian rocks. Subsequent deposition of platformal shales and carbonates continued through the Cambrian until the Ordovician, when quartz-rich and sandy clastics were again deposited. The Lower Palaeozoic succession of the Avalon Zone, which regionally is striking in its uniformity, records a relatively quiescent platformal environment that contrasts dramatically with the earlier history of the Avalon Zone, as well as with events recorded in synchronous rocks from the Appalachian mobile belt.

In the Devonian, the Avalon Zone was again the site of felsic magmatism and relatively restricted terrestrial sedimentation and volcanism. The region was affected by Acadian deformation and low grade regional metamorphism prior to the emplacement of major Devonian plutons (Dallmeyer *et al.*, 1983). Sedimentation and possibly magmatism continued into the Carboniferous (Bell and Blenkinsop, 1975).

9.11.7   *The Avalonian–Appalachian connection*

Avalonian rocks evolved mainly during the Grenville–Appalachian temporal gap, and are broadly coeval with and comparable to rocks formed in the later stages of the Pan-African orogenic episode (e.g., O'Brien *et al.*, 1983). Published models for the evolution of the zone are ambivalent, either favouring or discounting genetic links with the remainder of the Appalachian–Caledonian Orogen. New dates from latest Precambrian rift-related alkaline rocks from the western Avalon Zone (G. R. Dunning, pers. commun., 1989) are synchronous with the rift-drift transition recorded on the Appalachian miogeocline (Williams and Hiscott, 1987). These data, coupled with the presence of Pan-African suites within the orogen, may imply a broad genetic link between the latest stage Avalonian–Pan-African orogenic activity and earliest Palaeozoic initiation of the Appalachian Wilson cycle. Most geological, geochronological and geophysical evidence in southeast Newfoundland supports the hypothesis that the final accretion of the Avalon Zone to the remainder of the orogen was controlled by transcurrent movements perhaps as early as the Silurian (Kontak *et al.*, 1988) and through the Devonian. Deep seismic data support earlier notions, based in part on timing of magmatism and regional correlation, that the zone, as defined in Newfoundland (*cf.*, Williams, 1979) represents a distinct lithospheric block, whose final juxtaposition with inboard terranes of the Newfoundland Appalachians occurred in mid-Palaeozoic times (O'Brien *et al.*, 1983).

However, controls on timing of the earliest juxtaposition of Avalonian rocks with the remainder of the orogen are lacking. Metamorphic and protolith ages from gneisses and associated rocks exposed west of the Hermitage Bay Fault on the south coast of Newfoundland may indicate that higher grade and perhaps older events related to Avalonian tectonics are recorded within the Appalachian mobile belt in terranes inboard of the classic Avalon Zone (O'Brien *et al.*, 1983; Dunning and O'Brien, 1989). Because this crust lacks the diagnostic early Palaeozoic cover succession of the Avalon Zone, and are thus without clear 'zonal' affiliation, they have been excluded from this overview of the Avalon type area. Nevertheless, understanding the Precambrian history of these rocks, the time of their first Palaeozoic deformation, and the relation of the tectono-thermal history of these rocks to events recorded east of the Dover and Hermitage Bay faults are exciting, first order problems with major implications for Avalonian and Appalachian tectonics.

Acknowledgements

In preparing this chapter, the authors have summarized the work of many individuals. We apologise to those we may have inadvertently overlooked. We thank J. Tuach and E. Hussey for providing us with unpublished chemical data, S. Colman-Sadd and J. Hayes for assistance in the preparation of Figures 9.1 and 9.2, R. Stevens for drawing our attention to possible analogies between the Lord Howe Rise and the Avalon Zone, and C. O'Driscoll and P. O'Neill for critically reviewing the manuscript.

We offer sincere thanks to K. Byrne, G. Denief and T. Paltanavage (Geological Survey), who provided cartographic services, and to W. Marsh (Earth Sciences), who prepared the plates. Our work on Avalonian rocks has been funded, to varying degrees, by joint Canada–Newfoundland Mineral Development and related agreements. Research by D. F. S. and A. F. K. is also funded by the Natural Sciences and Engineering Research Council of Canada. S. J. O'B. publishes with the permission of the Head of the Geological Survey Branch.

References

Anderson, M. M. (1981) The Random Formation of southeastern Newfoundland: a discussion aimed at establishing its age and relationships to bounding formations. *Am. J. Sci.* **281**, 807–830.

Anderson, M. M. and Conway Morris, S. (1982) A review, with descriptions of four unusual forms of the soft-bodied fauna of the Conception and St. John's groups, Avalon Peninsula, Newfoundland. *Proc. Third N. Am. Paleonto. Conv.*

Anderson, M. M. and Misra, S. B. (1968) Fossils found in the Precambrian Conception Group. *Nature (London)* **220**, 680–681.

Anderson, M. M., Brueckner, W. D., King, A. F. and Maher, J. B. (1975) The late Proterozoic 'H. D. Lilly Unconformity' at Red Head, Northeast Avalon Peninsula, Newfoundland. *Am. J. Sci.* **275**, 1012–1027.

Atwater, T. (1970) Implication of plate tectonics for the Cenozoic evolution of western North America. *Bull. Geol. Soc. Am.* **81**, 3513–3536.

Bell, K. and Blenkinsop, J. (1975) Geochronology of eastern Newfoundland. *Nature (London)* **254**, 410–411.

Blackwood, R. F. (1977) Geology of the Hare Bay area, northwestern Bonavista Bay. In *Report of Activities for 1976*, ed. Gibbons, R., *Newfoundland Dept. Mines and Energy, Min. Dev. Div., Rept.* **77-5**, 7–14.

Blackwood, R. F. and Kennedy, M. J. (1975) The Dover Fault: Western boundary of the Avalon Zone in Newfoundland. *Can. J. Earth Sci.* **12**, 320–325.

Blackwood, R. F. and O'Driscoll, C. F. (1976) The Gander–Avalon boundary in northeast Newfoundland. *Can. J. Earth Sci.* **13**, 1155–1159.

Boyce, W. D. (1988) Cambrian trilobite faunas on the Avalon Peninsula, Newfoundland. *GAC-MAC-CSPG Field Trip Guidebook*, Trip **A-8**, 77 pp.

Bradley, D. A. (1962) Gisborne Lake and Terrenceville map-areas, Newfoundland. *Geol. Surv. Can. Mem.* **321**, 56 pp.

Butler, A. J. and Greene, B. A. (1976) Silica resources of Newfoundland. *Newfoundland Dept. Mines and Energy, Min. Dev. Div. Rept.* **76-2**, 68 pp.

Caron, A. (1986) Microstructural study of the Dover Fault, northeastern Newfoundland. In Program with Abstracts, *GAC-MAC-CGU Joint Annual Meeting*, Ottawa **11**, 52.

Caron, A. and Williams, P. (1988) The multistage development of the Dover Fault in northeastern Newfoundland: the late stages. In Program with Abstracts, *GAC-MAC-CSPG Annual Meeting*, St. John's **13**, A-17.

Cutt, B. J. and Laving, J. G. (1977) Tectonic elements and geologic history of the south Labrador and Newfoundland continental shelf, eastern Canada. *Bull. Can. Pet. Geol.* **25**, 1037–1058.

Dallmeyer, R. D. (1980) Geochronology Report. In *Current Research*, eds. O'Driscoll, C. F. and Gibbons, R. V., *Newfoundland Dept. of Mines and Energy, Min. Dev. Div. Rept.* **80-1**, 143–146.

Dallmeyer, R. D., Odom, A. L., O'Driscoll, C. F. and Hussey, E. M. (1981a) Geochronology of the Swift Current Granite and host volcanic rocks of the Love Cove Group, southwestern Avalon

Zone, Newfoundland: Evidence for a late Proterozoic volcanic–subvolcanic association. *Can. J. Earth Sci.* **18**, 699–707.

Dallmeyer, R. D., Blackwood, R. F. and Odom, A. L. (1981b) Age and origin of the Dover Fault: tectonic boundary between the Gander and Avalon Zones of the northeastern Newfoundland Appalachians. *Can. J. Earth Sci.* **18**, 1431–1442.

Dallmeyer, R. D., Hussey, E. M., O'Brien, S. J. and O'Driscoll, C. F. (1983) Chronology of tectono-thermal activity in the western Avalon Zone of the Newfoundland Appalachians. *Can. J. Earth Sci.* **20**, 355–363.

Dickson, W. L. (1983) Geology, geochemistry and mineral potential of the Ackley Granite and parts of the Northwest Brook and Eastern Meelpaeg complexes, southeast Newfoundland. *Newfoundland Dept. Mines and Energy, Min. Dev. Div. Rept.* **83-6**, 129 pp.

Dunning, G. R. and O'Brien, S. J. (1989) Hadrynian gneiss and related rocks in the Appalachian mobile belt of southern Newfoundland: U/Pb ages and tectonic significance. *Geology* **17**, 548–551.

Durling, P. W., Bell, J. S. and Fader, G. B. J. (1987) The geological structure and distribution of Palaeozoic rocks on the Avalon Platform, offshore Newfoundland. *Can. J. Earth Sci.* **24**, 1412–1420.

Fletcher, T. P. (1972) Geology and Lower to Middle Cambrian trilobite faunas of the southwest Avalon, Newfoundland. Unpublished Ph.D. thesis, Cambridge University, Cambridge, England, 530 pp.

Fryer, B. J., Dickson, W. L., O'Brien, S. J., Wilton, D. C., Jenner, G. A. and Longstaffe, F. J. (1988) Geochemistry and tectonic significance of Newfoundland granitoids. *Lithoprobe East: Report of Transect Meeting*, St. John's, Newfoundland 30–33.

Gardner, S. and Hiscott, R. N. (1988) Deep-water facies and depositional setting of the lower Conception Group (Hadrynian), southern Avalon Peninsula, St. John's, Newfoundland. *Can. J. Earth Sci.* **25**, 1579–1594.

Greene, B. A. and O'Driscoll, C. F. (1976) Gaultois map area. In Report of Activities. *Newfoundland Dept. Mines and Energy, Min. Dev. Div. Rept.* **76-1**, 56–63.

Greene, B. A. and Williams, H. (1974) New fossil localities and the base of the Cambrian in southeast Newfoundland. *Can. J. Earth Sci.* **11**, 319–323.

Greenough, J. D. (1984) Petrology and geochemistry of Cambrian volcanic rocks from the Avalon Zone in eastern Newfoundland and southern New Brunswick. Ph.D. thesis, Department of Earth Sciences, Memorial University of Newfoundland, 487 pp.

Greenough, J. T. and Papezik, V. S. (1985) Petrology and geochemistry of Cambrian volcanic rocks from the Avalon Peninsula, Newfoundland. *Can. J. Earth Sci.* **22**, 1594–1601.

Haworth, R. T. and Lefort, J. P. (1979) Geophysical evidence for the extent of the Avalon Zone in Atlantic Canada. *Can. J. Earth Sci.* **16**, 552–567.

Hayes, A. O. (1948) Geology of the area between Bonavista and Trinity Bays, eastern Newfoundland. *Newfoundland Geol. Surv. Bull.* **32**(1), 1–37.

Henningsmoen, G. (1969) Short account of Cambrian and Tremadocian of Acado-Baltic province. In *North Atlantic—Geology and Continental Drift*, ed. Kay, M., *Am. Assoc. Pet. Geol. Mem.* **12**, 110–114.

Hiscott, R. N. (1982) Tidal deposits of the Lower Cambrian Random Formation, eastern Newfoundland: facies and paleoenvironments. *Can. J. Earth Sci.* **19**, 2028–2042.

Hodych, J. P. and Hayatsu, A. (1980) K–Ar isochron age and palaeomagnetism of diabase along the trans-Avalon aeromagnetic lineament—evidence of late Triassic rifting in Newfoundland. *Can. J. Earth Sci.* **17**, 491–499.

Hofmann, H. S., Hill, J. and King, A. F. (1979) Late Precambrian microfossils, southeast Newfoundland. *Geol. Surv. Can. Rept.* **79-1B**, 83–88.

Howell, B. F. (1925) The fauna of the Paradoxides beds at Manuels, Newfoundland. *Bull. Am. Paleo.* **11**(43), 140 pp.

Hughes, C. F. (1970) The late Precambrian Avalonian Orogeny in Avalon, southeast Newfoundland. *Am. J. Sci.* **269**, 183–190.

Hussey, E. M. (1979) Geology of the Clode Sound area, Newfoundland. Unpublished M.Sc. thesis, Memorial University of Newfoundland, St. John's, Newfoundland, 312 pp.

Hussong, D. M. and Uyeda, S. (1981) Tectonics in the Mariana arc: results of recent studies including DSDP Leg 60. *Oceanol. Acta* 203–211.

Hutchinson, R. D. (1962) Cambrian stratigraphy and trilobite faunas of southeastern Newfoundland. *Geol. Surv. Can. Bull.* **88**, 156 pp.

Jenness, S. E. (1963) Terra Nova and Bonavista map areas, Newfoundland. *Geol. Surv. Can. Mem.* **327**, 184 pp.

Jongsma, D. and Mutter, J. C. (1978) Non-axial rifting of a rift valley: evidence from the Lord Howe Rise and the southeast Australian margin. *Earth Planet. Sci. Lett.* **39**, 226–234.

Kay, M. and Colbert, E. H. (1965) *Stratigraphy and Life History*. Wiley, New York, 736 pp.

Keen, C. E., Keen, M. J., Nichols, B., Reid, I., Stockmal, G. S., Colman-Sadd, S. P., O'Brien, S. J., Miller, H., Quinlan, G., Williams, H. and Wright, J. (1986) Deep seismic reflection profiling across the northern Appalachians. *Geology* **14**, 141–145.

Kennedy, M. J., Blackwood, R. F., Colman-Sadd, S. P., O'Driscoll, C. F. and Dickson, W. L. (1982) The Dover–Hermitage Bay Fault: Boundary between the Gander and Avalon Zones, eastern Newfoundland. In *Major Structural Zones and Faults of the Northern Appalachians*, eds. St. Julien, P. and Beland, J., *GAC Spec. Publ.* **24**, 231–248.

King, A. F. (1980) The birth of the Caledonides: late Precambrian rocks of the Avalon Peninsula, Newfoundland and their correlatives in the Appalachian–Caledonian orogen. In *The Caledonides in the U.S.A.*, ed. Wones, D. R., Dept. Geol. Sci. VPI and SU Memoir **2**, 3–8.

King, A. F. (1986) Geology of the St. John's area, Newfoundland. In *Current Research*, eds. Blackwood, R. F., Walsh, D. G. and Gibbons, R. V., *Newfoundland Dept. Mines and Energy, Min. Dev. Div. Rept.* **86-1**, 209–218.

King, A. F. (1988) Geology of the Avalon Peninsula, Newfoundland (parts of 1K, 1L, 1M, 1N and 2C). *Newfoundland Dept. of Mines, Min. Dev. Div. Map* **88-01**.

King, A. F. (1990) Geology of the St. John's area. *Newfoundland Dept. Mines and Energy, Min. Dev. Div. Rept.* (in press).

King, L. D., Faber, G. B. J., Jenkins, W. A. M. and King, E. L. (1986) Occurrence and regional geological setting of Paleozoic rocks on the Grand Banks of Newfoundland. *Can. J. Earth Sci.* **23**, 504–526.

King, L. H., Fader, G. B., Poole, W. H. and Wanless, R. K. (1985) Geological setting and age of the Flemish Cap granodiorite, east of the Grand Banks of Newfoundland. *Can. J. Earth Sci.* **22**, 1286–1298.

Knight, I. and O'Brien, S. J. (1988) Stratigraphy and sedimentological studies of the Connecting Point Group, portions of the Eastport (2C/12) and St. Brendans (2C/13) map areas, Bonavista Bay, Newfoundland. In *Current Research*, eds. Hyde, R. S., Walsh, D. and Blackwood, R. F., *Newfoundland Dept. Mines and Energy, Min. Dev. Div. Rept.* **88-1**, 207–228.

Kontak, D. J., Tuach, J., Strong, D. F., Archibald, D. A. and Farrar, E. (1988) Plutonic and hydrothermal events in the Ackley Granite, southeast Newfoundland as recorded by total fusion $^{40}Ar/^{39}Ar$ geochronology. *Can. J. Earth Sci.* **25**, 1151–1160.

Krogh, T. E., Strong, D. F., O'Brien, S. J. and Papezik, V. S. (1988) Precise U–Pb age dates of zircons from the Avalon Terrane in Newfoundland. *Can. J. Earth Sci.* **25**, 442–453.

Kroner, A. (1980) Pan-African crustal evolution. *Episodes* **2**, 3–8.

Leblanc, M. (1981) The late Proterozoic ophiolites at Bou Azzer, Morocco: evidence for Pan-African plate tectonics. In *Precambrian Plate Tectonics*, ed. Kroner, A. Elsevier, Amsterdam, 435–441.

Lefort, J. P., Max, M. D. and Roussel, J. (1988) Geophysical evidence for the location of the NW boundary of Gondwanaland and its relationship to two older satellite sutures. In *The Caledonian–Appalachian Orogen*, eds. Harris, A. L. and Fettes, D. J., *Geol. Soc. Spec. Publ.* **38**, 49–60.

Lilly, H. D. (1966) Late Precambrian and Appalachian tectonics in light of submarine exploration of the Great Bank of Newfoundland in the Gulf of St. Lawrence. *Am. J. Sci.* **264**, 569–574.

Lipman, P. W., Prostka, H. J. and Christiansen, R. L. (1972) Cenozoic volcanism and plate tectonic evolution of the western United States. *Philos. Trans. R. Soc. London Ser. A* **271**, 217–284.

Marillier, F., Keen, C. E., Stockmal, G. S., Quinlan, G., Williams, H., Colman-Sadd, S. P. and O'Brien, S. J. (1989) Crustal structure and surface zonation of the Canadian Appalachians: Implications of deep seismic reflection data. *Can. J. Earth Sci.* **26**, 305–321.

Martin, F. and Dean, W. T. (1981) Middle and Upper Cambrian and Lower Ordovician acritarchs from Random Island, eastern Newfoundland. *Geol. Surv. Can. Bull.* **343**, 43 pp.

McCartney, W. D. (1967) Whitbourne map-area, Newfoundland. *Geol. Surv. Can. Mem.* **341**, 133 pp.

Miller, H. (1977) Gravity zoning in Newfoundland. *Tectonophysics* **38**, 316–326.

Narbonne, G. M., Myrow, P., Landing, E. and Anderson, M. M. (1987) A candidate stratotype for the Precambrian–Cambrian boundary, Fortune Head, Burin Peninsula, southeast Newfoundland. *Can. J. Earth Sci.* **24**, 1277–1293.

Nixon, G. T. and Papezik, V. S. (1979) Late Precambrian ash-flow tuffs and associated rocks of the Harbour Main Group near Colliers, eastern Newfoundland: chemistry and magmatic affinities. *Can. J. Earth Sci.* **16**, 167–181.

O'Brien, S. J. (1987) Geology of the Eastport (west half) map area, Bonavista Bay, Newfoundland. In *Current Research*, eds. Walsh, D., Blackwood, R. F. and Gibbons, R. V., *Newfoundland Dept. Mines and Energy, Min. Dev. Div. Rept.* **87-1**, 257–270.

O'Brien, S. J. and Knight, I. (1988) The Avalonian geology of southwest Bonavista Bay: portions of the St. Brendan's (2C/13) and Eastport (2C/12) map areas. In *Current Research*, eds. Hyde, R. S., Walsh, D. and Blackwood, R. F., *Newfoundland Dept. Mines and Energy, Min. Dev. Div. Rept.* **88-1**, 193–206.

O'Brien, S. J. and Taylor, S. W. (1983) Geology of the Baine Harbour (1M/7) and Point Enragee (1M/6) map areas, southeast Newfoundland. *Newfoundland Dept. Mines and Energy, Min. Dev. Div. Rept.* **83-5**, 70 pp.

O'Brien, S. J., Strong, P. G. and Evans, J. L. (1977) Geology of the Grand Bank (1M/4) and Lamaline (1L/13) map areas. *Newfoundland Dept. Mines and Energy, Min. Dev. Div. Rept.* **77-7**, 16 pp.

O'Brien, S. J., O'Driscoll, C. F. and Colman-Sadd, S. P. (1981) Geological compilation of Belleoram (1M) and St. Lawrence (1L), Newfoundland. *Newfoundland Dept. Mines and Energy, Min. Dev. Div. Map* **81-116**.

O'Brien, S. J., Wardle, R. J. and King, A. F. (1983) The Avalon Zone: A Pan-African terrane in the Appalachian Orogen of Canada. *Geol. J.* **18**, 195–222.

O'Brien, S. J., Nunn, G. A. G., Dickson, W. L. and Tuach, J. (1984) Geology of the Terrenceville (1M/10) and Gisborne Lake (1M/15) map areas, southeast Newfoundland. *Newfoundland Dept. Mines and Energy, Min. Dev. Div. Rept.* **84-4**, 54 pp.

O'Brien, S. J., Strong, D. F. and Dostal, J. (1986) Stratigraphic and petrochemical evolution of late Proterozoic rocks in southeastern Newfoundland. *Mar. Sed. Atlantic Geol.* **22**, 196–197.

O'Driscoll, C. F. (1977) Geology, petrology and geochemistry of the Hermitage Peninsula, southern Newfoundland. Unpublished M.Sc. thesis, Memorial University of Newfoundland, St. John's Newfoundland, 144 pp.

O'Driscoll, C. F. and Strong, D. F. (1978) Geology and geochemistry of late Precambrian volcanic and intrusive rocks of southwestern Avalon Zone in Newfoundland. *Precambrian Res.* **8**, 19–48.

Papezik, V. S. and Barr, S. M. (1981) The Shelbourne Dyke, an early Mesozoic diabase dyke in Nova Scotia: mineralogy, chemistry and regional significance. *Can. J. Earth Sci.* **18**, 1346–1355.

Pelletier, B. (1971) A granodiorite drill core from the Flemish Cap, eastern Canadian continental shelf. *Can. J. Earth Sci.* **8**, 1499–1503.

Poulsen, V. and Anderson, M. M. (1975) The Middle–Upper Cambrian transition in southeastern Newfoundland. *Can. J. Earth Sci.* **12**, 1710–1726.

Ranger, M. J., Pickerill, R. K. and Fillion, D. (1984) Lithostratigraphy of the Cambrian?–Lower Ordovician Bell Island and Wabana groups of Bell, Little Bell and Kelly's islands, Conception Bay, eastern Newfoundland. *Can. J. Earth Sci.* **21**, 1245–1261.

Rodgers, J. (1967) Chronology of tectonic movements in the Appalachian region of eastern North America. *Am. J. Sci.* **265**, 408–427.

Rodgers, J. (1972) Late Precambrian (post-Grenville) rocks of the Appalachian region. *Am. J. Sci.* **272**, 507–520.

Rose, E. R. (1952) Torbay map area, Newfoundland. *Geol. Surv. Can. Mem.* **265**, 64 pp.

Sheridan, R. E. and Drake, C. L. (1968) Seaward extension of the Canadian Appalachians. *Can. J. Earth Sci.* **5**, 337–373.

Smith, S. A. and Hiscott, R. N. (1984) Latest Precambrian to Early Cambrian basin evolution, Fortune Bay, Newfoundland: fault-bounded basin to platform. *Can. J. Earth Sci.* **21**, 1379–1392.

Stern, R. J., Gottfried, D. and Hedge, C. E. (1984) Late Precambrian rifting and crustal evolution in the Northeastern Desert of Egypt. *Geology* **12**, 168–172.

194   AVALONIAN AND CADOMIAN GEOLOGY OF THE NORTH ATLANTIC

Strong, D. F. (1979) Proterozoic tectonics of Northwestern Gondwanaland: New evidence from eastern Newfoundland. *Tectonophysics* **54**, 81–101.

Strong, D. F. and Dostal, J. (1980) Dynamic partial melting of Proterozoic upper mantle: Evidence from rare earth elements in oceanic crust of eastern Newfoundland. *Contrib. Mineral. Petrol.* **72**, 165–173.

Strong, D. F., O'Brien, S. J., Strong, P. G., Taylor, S. W. and Wilton, D. H. C. (1978) Aborted Proterozoic rifting in Newfoundland. *Can. J. Earth Sci.* **15**, 117–131.

Taylor, S. W. (1976) Geology of the Marystown map sheet (E/2), Burin Peninsula, southeastern Newfoundland. Unpublished M.Sc. thesis, Memorial University of Newfoundland, St. John's, Newfoundland, 164 pp.

Uyeda, S. and Miyashiro, A. (1974) Plate tectonics and the Japanese islands: a synthesis. *Bull. Geol. Soc. Am.* **85**, 1159–1170.

Van Ingen, G. (1914) Table of the geological formation of the Cambrian and Ordovician systems about Conception and Trinity bays, Newfoundland and their North American and western European equivalents, based on 1912–1913 fieldwork. *Princeton University Contributions* **4**.

Walcott, C. D. (1900) Random, a pre-Algonquin terrane. *Bull. Geol. Soc. Am.* **11**, 4–5.

Wernicke, B. (1985) Uniform-sense normal simple shear of the continental lithosphere. *Can. J. Earth Sci.* **22**, 108–125.

White, D. E. (1939) Geology and molybdenite deposits of the Rencontre East area, Fortune Bay, Newfoundland. Unpublished Ph.D. thesis, Princeton University, Princeton, New Jersey, 119 pp.

Widmer, K. (1950) The geology of the Hermitage Bay area, Newfoundland. Unpublished Ph.D. thesis, Princeton University, Princeton, New Jersey, 439 pp.

Williams, H. (1964) The Appalachians in Newfoundland—a two-sided symmetrical system. *Am. J. Sci.* **262**, 1137–1158.

Williams, H. (1971) Geology of the Belleoram map area, Newfoundland. *Geol. Surv. Can. Pap.* **70-65**, 39 pp.

Williams, H. (1979) Appalachian Orogen in Canada. *Can. J. Earth Sci.* **16**, 792–807.

Williams, H. and Hatcher, R. D., Jr. (1983) Appalachian suspect terranes. In *Contributions to the Tectonics of Mountain Chains*, eds. Hatcher, R. D., Jr., Williams, H. and Zietz, I., *Geol. Soc. Am. Mem.* **158**, 33–53.

Williams, H. and Hiscott, R. N. (1987) Definition of the Iapetus rift–drift transition in western Newfoundland. *Geology* **15**, 1044–1047.

Williams, H. and King, A. F. (1979) Trepassey map area, Newfoundland. *Geol. Surv. Can. Mem.* **389**, 24 pp.

Williams, H., Kennedy, M. J. and Neale, E. R. W. (1972) The Appalachian Structural Province. In *Variations in Tectonic Styles in Canada*, eds. Price, R. A. and Douglas, R. J. W., *GAC Spec. Pap.* **11**, 181–261.

Wilson, J. T. (1988) Convection tectonics: some possible effects upon the earth's surface of flow from the deep mantle. *Can. J. Earth Sci.* **25**, 1199–1208.

Winchester, J. A. and Floyd, P. A. (1977) Geochemical discrimination of different magma series and their differentiation products using immobile elements. *Chem. Geol.* **20**, 325–343.

Younce, G. B. (1970) Structural geology and stratigraphy of the Bonavista Bay region, Newfoundland. Unpubl. Ph.D. dissertation, Cornell University, Ithica, New York, 188 pp.

Zeitz, I., Haworth, R. T., Williams, H. and Daniels, D. L. (1980) Magnetic anomaly map of the Appalachian Orogen. *Memorial University of Newfoundland, Map* **2**, Scale: 1:1 000 000.

10 The Avalon composite terrane of Nova Scotia

J. B. MURPHY, J. D. KEPPIE, R. D. NANCE and J. DOSTAL

10.1 Introduction

The Avalon composite terrane is defined by the presence of a lithostrati-graphically correlative, subaerial to shallow marine, Cambro-Ordovician overstep sequence containing an Acado-Baltic fauna (Keppie, 1985). In general, these Cambro-Ordovician rocks are only preserved in outliers, so overstep relationships are inferred from their close lithostratigraphic similarity. Where the Cambro-Ordovician rocks have been removed by erosion, it is possible to use the presence of the lithostratigraphically correlative, Silurian-Gedinnian overstep sequence containing the distinctive Rhenish-Bohemian fauna (Boucot, 1975). By this definition, the Avalon composite terrane is an early Palaeozoic entity and includes most, if not all, of northern Nova Scotia (Figures 10.1–10.3). In these areas, the early Palaeozoic rocks rest unconformably-disconformably-conformably upon a variety of late Precambrian volcanic rocks and turbidites which in turn overlie Middle to Late Proterozoic or older metasediments and gneisses. Contrasts in the Precambrian stratigraphy initially led to the subdivision of the Avalon composite terrane into several Late Proterozoic terranes (Keppie, 1985, 1989). The main terranes were: (a) Cape Breton terrane: a cratonic magmatic arc; (b) Antigonish Highlands and Cheticamp terranes: inter-arc basins floored by oceanic lithosphere; and (c) Cobequid Highlands terrane: a cratonic magmatic arc. However, new data has allowed lithostratigraphic correlations to be made across northern Nova Scotia which suggest the presence of just one terrane consisting of alternating basins floored by thinned continental lithosphere (rather than oceanic lithosphere) and horsts (Keppie et al., 1990a). This terrane is interpreted as a magmatic arc–intra-arc rift complex that lacks oceanic components. The relatively continuous exposure of Precambrian rocks in this region make it reasonably certain that no major sutures occur within it.

In contrast, Barr and Raeside (1986, 1989) have subdivided Cape Breton Island into four, apparently contrasting, tectonostratigraphic zones or ter-ranes, designated from southeast to northwest: Southeastern, Bras d'Or, Highlands and Northwestern Highlands. They define these terranes as follows: (1) Southeastern (Mira) terrane: late Precambrian volcanic and

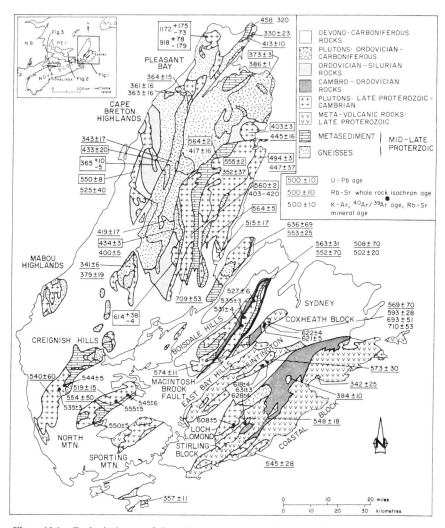

Figure 10.1 Geological map of Cape Breton Island (modified after Barr and Raeside, 1986) showing published geochronological data. References for these data may be obtained in Keppie *et al.* (1990b). Inset shows locations of Figures 10.1, 10.2 and 10.3. Solid circles indicate locations of ^{40}Ar/^{39}Ar data.

plutonic rocks overlain by Cambro-Ordovician rift basin sedimentary and minor volcanic rocks; (2) Bras d'Or terrane: gneissic basement overlain by platformal sedimentary rocks (George River Group) intruded by mainly late Precambrian and Ordovician (?) granitoid rocks: (3) Highlands terrane: gneissic core flanked by low grade sedimentary and volcanic rocks of probable Precambrian age intruded by diverse Precambrian–Carboniferous plutons: and (4) North-western Highlands terrane: gneissic basement intruded

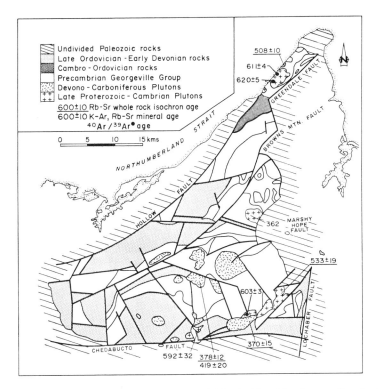

Figure 10.2   Geological map of the Antigonish Highlands (after Murphy and Keppie, 1987) showing published geochronological data. References for these data may be obtained in Keppie *et al.* (1990b). Location is shown in the inset to Figure 10.1. Solid circles indicate location of ^{40}Ar/^{39}Ar data.

by varied plutonic rocks including anorthosite and Grenvillian syenite. Barr and Raeside (1986) inferred that the Southeastern and Bras d'Or terranes are part of the Avalon terrane. Subsequently, Raeside (1987) split the Highlands terrane into the Aspy and Ingonish terranes and then incorporated the Ingonish terrane within the Bras d'Or terrane (Loncarevic *et al.*, 1989).

Several conflicting correlations between Cape Breton Island and Newfoundland have been proposed. The Bras d'Or terrane has been correlated with the Avalon Zone (Barr and Raeside, 1986; Marillier *et àl.*, 1989) or the Gander Zone (Barr *et al.*, 1987; Loncarevic *et al.*, 1989; Stockmal *et al.*, 1989). The Highlands/Aspy terrane was correlated with (1) the Gander terrane or outboard edge of the Piedmont terrane in Newfoundland (Barr and Raeside, 1986); (2) the Gander zone and Exploits subzone (eastern Dunnage zone), by Loncarevic *et al.* (1989) and (3) the Gander Zone by Stockmal *et al.* (1989). The Northwestern Highlands terrane was inferred to be an outboard part of autochthonous, Grenvillian North America (Piedmont terrane of Williams and Hatcher, 1982; North American terrane of Keppie, 1988).

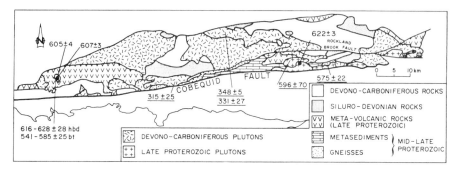

Figure 10.3   Geological map of the Cobequid Highlands showing published geochronological data. References for these data may be obtained in Keppie *et al.* (1990b). Location is shown in the inset to Figure 10.1. Solid circles indicate location of ^{40}Ar/^{39}Ar data.

However, other data suggest that Cape Breton Island, like mainland northern Nova Scotia, represents a single Avalonian terrane. Keppie (1979, 1982) and Currie (1987) have shown that the felsic and mafic gneisses in the Northwestern Highlands 'terrane' may be traced through both the Highlands and Bras d'Or 'terranes'. This is supported by sulphur and lead isotopic values from the carbonate-hosted Meat Cove and Lime Hill zinc deposits which have typical Grenvillian signatures (Sangster and Thorpe, 1988). In addition, the Precambrian platformal unit characterized by the George River Group (Keppie, 1979) in the Bras d'Or 'terrane' may be traced into the Highlands 'terrane' as the McMillan Flowage Formation (of Barr and Raeside, 1986) and Cape North Group (of MacDonald and Smith, 1980). Also, Late Proterozoic volcanic rocks and the Cambro-Ordovician overstep sequence are present in both Southeastern and Bras d'Or 'terranes'. Recently, pebbles containing Cambro-Ordovician and Silurian fauna typical of the Avalon composite terrane have been found in a Lower Carboniferous conglomerate derived from the Mabou Highlands (Keppie *et al.*, 1990a) lying within the Highlands 'terrane'. Furthermore, published geochronological data show that Late Proterozoic–Cambrian and Palaeozoic plutons are present across most of Cape Breton Island. The basement beneath the Southeastern (Mira) 'terrane' is not exposed. However, the geochemistry of the Late Proterozoic volcanic rocks indicates the presence of continental crust (Keppie *et al.*, 1979). Thus, these data suggest that the rocks of Cape Breton Island, rather than representing an entire cross-section of the Appalachian Orogen and distinct terranes as proposed by Barr and Raeside (1986), may expose an oblique section of the Avalon composite terrane with progressively deeper levels of the Late Proterozoic magmatic arc complex cropping out northwestwards. That the gneisses in northern Cape Breton Island are not contiguous with the outer edge of cratonic North American basement is indicated on seismic reflection profiles between Newfoundland and Cape Breton Island which shows that the top of the Grenvillian basement is at a depth of >10 s

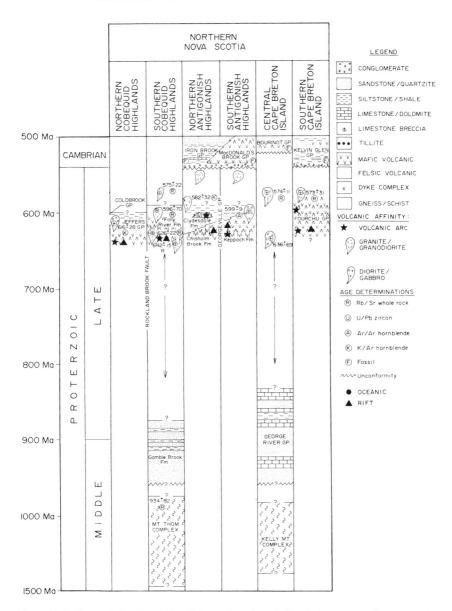

Figure 10.4   Precambrian–Cambrian lithostratigraphy of the Avalon composite terrane in Nova Scotia. References for these data may be obtained in Kappie *et al.* (1990b).

two-way travel time in the Cabot Strait off northern Cape Breton Island (Marillier *et al.*, 1989). Resolution of these alternative hypotheses and of the geological history may be achieved by supplementing the geological data base with further age dating.

10.2   Geological data

10.2.1   Mid/Late Proterozoic gneisses

The oldest rocks in the Avalon composite terrane in Nova Scotia are thought to be represented by ortho- and para-gneisses and amphibolite sheets, although reliable geochronological data supporting their antiquity are sparse. These rocks comprise the Kellys Mountain and Pleasant Bay Complexes (includes the Polletts Cove Brook Group of Barr and Raeside, 1986) and equivalents in Cape Breton Island (Figures 10.1 and 10.4), the Great Village River Gneiss and Mount Thom Complex in the Cobequid Highlands of Nova Scotia (Figures 10.3 and 10.4).

Gneisses of the Mount Thom Complex (Figure 10.3) exhibit polyphase deformation consisting of an early metamorphic foliation folded by isoclinal folds associated with an axial planar foliation and amphibolite facies metamorphism. This, in turn, is deformed by a crenulation cleavage (Donohoe, 1976; Donohoe and Cullen, 1983).

Farther west in the Cobequid Highlands, the Great Village River Gneiss is composed of hornblende orthogneisses, biotite-garnet psammitic paragneisses and quartz-plagioclase layered amphibolite dykes. Doig et al. (1989) obtained a 734 ± 2 Ma U–Pb age for the orthogneisses on Economy River. (The age of the remainder of the Great Village River Gneiss is currently under investigation.) They may represent either a pre-Avalonian, i.e. pre-650 Ma basement, or the metamorphic infrastructure of a late Precambrian magmatic arc. These gneisses are reported to possess early metamorphic foliation developed under amphibolite facies metamorphism (Cullen, 1984), which is isoclinically folded and is truncated by a shear zone forming the boundary of the Gamble Brook Formation (Murphy et al., 1988). The earliest metamorphic fabric (S1) is not present in the Gamble Brook Formation, which may suggest that the two units were separated by an unconformity. However, the present contact between these units is a ductile shear zone and the presence of an unconformity cannot be confirmed. It is also possible that S1 records the earliest phase of progressive deformation which also deforms the Gamble Brook Formation. Syntectonic granite gneisses intrusive into the Great Village Gneiss have yielded c. 610–605 Ma U–Pb zircon ages (Doig et al., 1989).

In Cape Breton Island (Figure 10.1) the Kellys Mountain Complex is composed of both ortho- and para-gneisses (biotite gneiss, cordierite gneiss, quartzo-feldspathic gneiss and migmatite) cut by amphibolite sheets (Keppie, 1982, 1985; Jamieson, 1984). The gneisses display folds in the gneissic banding with a parallel foliation accompanied by amphibolite facies metamorphism attributed to the Micmacian Orogeny by Keppie (1982) (Figure 10.5). These rocks are intruded by a variety of plutonic rocks including diorite and granite (Barr et al., 1982), which may be related to the low pressure, upper amphibolite facies metamorphism including anatexis (Jamieson, 1984). $^{40}Ar/^{39}Ar$ analyses of hornblende in the gneisses, amphibolites and diorites yielded *c.* 500–490 Ma plateau ages interpreted to closely post-date the intrusion of the Kellys Mountain granite (Keppie and Dallmeyer, 1989). The gneisses have given a 701 ± 66 Ma Rb–Sr whole-rock isochron which may

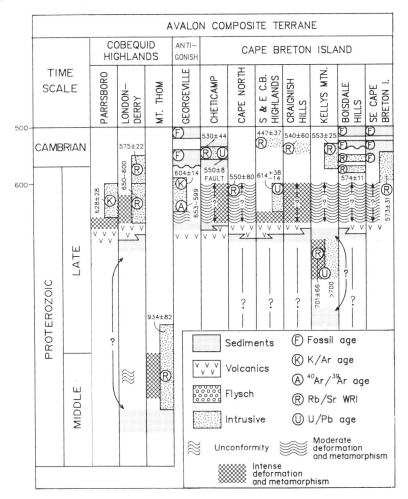

Figure 10.5 Constraints on the time of deformation and metamorphism in the Avalon composite terrane in the northern Appalachians. Parrsboro: polyphase deformation at greenschist facies metamorphism in the Jeffers Group is cut by the Jeffers Brook pluton dated at 628 ± 22 Ma (K–Ar on hornblende). Londonderry: polyphase deformation at greenschist facies metamorphism in the Folly River Formation is cut by the McCallum Settlement pluton (575 ± 22 Ma, Rb–Sr whole-rock isochron and the Debert River pluton 596 ± 70 Ma, Rb–Sr whole-rock isochron). Georgeville Group cut by Greendale pluton dated at 604 ± 14 (K–Ar on muscovite) and 653–599 ± 9 Ma (^{40}Ar/^{39}Ar on hornblende) unconformably overlain by Early Cambrian rocks. Boisdale Hills: polyphase deformation at greenschist facies metamorphism cut by Shunacadie and Boisdale Hills plutons dated at 574 ± 11 Ma and 563 ± 31 Ma. Southeastern Cape Breton Island: single phase of deformation at greenschist facies metamorphism in Fourchu Group cut by plutons dated between 573 ± 30 and 548 ± 18 Ma (Rb–Sr whole-rock isochrons) and overlain by Cambrian rocks containing pebbles with a pre-depositional fabric. References for these data may be obtained in Keppie *et al.* (1990b).

provide a lower limit on the time of the early metamorphism (Olszewski *et al.,* 1981).

The Pleasant Bay Complex underlies large areas in the western Cape Breton Highlands and is made up of biotite gneiss, amphibolitic gneiss, gneissic granodiorite, marble, quartzite and schist cut by lit-par-lit gneiss (Currie, 1987). The gneisses display polyphase deformation and several episodes of metamorphism: the early metamorphism is largely obscured by a high pressure and high temperature upper amphibolite (-granulite) facies metamorphism overprinted by late retrograde metamorphism. A syenite intrusive into gneisses of the Pollets Cove Brook Group (= Pleasant Bay Complex) in northern Cape Breton Island has yielded an U–Pb chord with a lower intercept age of $918 + 78 - 179$ Ma and an upper intercept age of $1172 + 135 - 73$ Ma (Barr *et al.,* 1987). Using these data, Barr *et al.* (1987) estimated the age of amphibolite metamorphism to be 1045–1040 Ma although it is not clear how this estimate was derived. An alternative interpretation would have the upper intercept as the time of intrusion and the lower intercept as the time of metamorphism. The abraded suite of zircons lies towards the top end of the chord suggesting that it is projecting towards the time of intrusion, while the euhedral shape of the core is consistent with an igneous origin. The lower intercept would then represent a minimum age for the high grade metamorphsim and deformation.

While more geochronological data are essential, existing data in the

Figure 10.6   Tectonically interleaved marble (white) and pelitic gneisses (grey) of the Cape North Group, Cape Breton Island.

Figure 10.7 Folded mylonitic fabric developed in the Gamble Brook Formation within the ductile shear contact zone between the Great Village River Gneiss and the Gamble Brook Formation, Cobequid Highlands.

Avalon composite terrane tentatively suggest that a metamorphic event took place around the Middle-Late Proterozoic boundary.

10.2.2 Mid/Late Proterozoic platformal rocks

The next lithostratigraphic unit (Figures 10.1, 10.3 and 10.4) is represented by strongly deformed, miogeoclinal metasedimentary rocks: quartzites, carbonates, greywackes and slates (metavolcanic rocks) of the McMillan Flowage Formation and George River and Cape North Groups (Cape Breton Island, Figure 10.6), and Gamble Brook Formation (Cobequid Highlands, Figure 10.7). The age of these units is based upon their lithostratigraphic correlation with the Green Head Group in southern New Brunswick which contain stromatolites assigned a Middle Riphean age by Hoffman (1974). This would tentatively correspond to $c. 1000 \pm 200$ Ma. In most places the contact between the gneisses and the metasediments is tectonic. These platformal units generally display polyphase deformation and the metamorphism ranges from greenschist facies in southern Cape Breton Island and the Cobequid Highlands to amphibolite facies in northern Cape Breton Island (Keppie, 1979). The tectonic nature of the contact between these platformal units and the Late Proterozoic sequences has led to much divergence of opinion on the ages of these structures: pre-Late Proterozoic,

Figure 10.8   Concordant syntectonic granite gneiss (centre) that occurs within the ductile shear contact zone between the Great Village River Gneiss and the Gamble Brook Formation, Cobequid Highlands.

post-Late Proterozoic/pre-Eocambrian or Palaeozoic. In most places the data is equivocal. However, in the eastern Cobequid Highlands, intense mylonitic fabrics in the Gamble Brook Formation are truncated by a mafic dyke intruded along the contact with the Late Proterozoic Folly River Formation. This mafic dyke is petrographically and chemically similar to the mafic lavas in the Folly River Formation and is inferred to be a feeder dyke. This suggests that the contact between the Gamble Brook, and Folly River formations is an unconformity and that deformation took place between their deposition. Syntectonic granite gneisses in the mylonitic contact zone (Figure 10.8) between the Great Village River Gneiss and Gamble Brook Formation have yielded c. 610–605 Ma ages (U–Pb zircon, Doig et al., 1989). The above data indicate that the time of deformation may span c. 630–605 Ma. C-S fabrics, gently east-southeast plunging stretching lineations defined by dimensionally oriented hornblende, quartz and quartz-feldspar augen, and fold asymmetry in a moderately to steeply southeast-dipping mylonitic fabric indicate an oblique sense of shear with both normal and sinistral components suggesting that the mylonite zone was developed in a transtensional environment (Nance and Murphy, 1990).

10.2.3   Late Proterozoic volcano-sedimentary rocks

The rest of the Late Proterozoic rocks of the Avalon composite terrane are characterized by volcanic rocks and clastic metasediments (Figures 10.1–10.4). These latest Proterozoic rocks are exposed in southern Cape Breton Island (Fourchu Group and equivalents, Figure 10.1), in the Antigonish Highlands (Georgeville Group, Figure 10.2), in the northern Cobequid Highlands (Jeffers Group = Warwick Mountain Formation, Figure 10.3) and in the southern Cobequid Highlands (Folly River Formation).

In southern Cape Breton Island, the Fourchu Group and its equivalents occur in several fault-bounded blocks (Figure 10.1). The Fourchu Group and its correlatives are primarily composed of subaerial–shallow marine pyroclastic rocks (crystal tuff, ignimbrite, breccia, lithic tuff and ash) and bimodal, volcanic arc flows (Keppie et al., 1979, 1990b). The base of the Fourchu Group is not exposed. However, the geochemistry of the Fourchu Group indicates the presence of thinned continental crust beneath the Fourchu Group during the Late Proterozoic. Helmstaedt and Tella (1973) inferred that volcanic pebbles near the top of the George River Group were lateral equivalents of part of the Fourchu Group. This suggests that the Fourchu Group may be underlain by the George River Group and its basement. The Fourchu Group and its correlatives are associated with hornblende-biotite, I-type, epizonal, calc-alkaline diorites, tonalites and granites (Barr et al., 1982) which give ages in the range c. 615–515 Ma ($^{40}Ar/^{39}Ar$ on hornblende; Keppie et al., 1990b). Contact metamorphic minerals indicate a generally low pressure

Figure 10.9   Sub-horizontal, interlayered mudstones (dark) and greywackes (pale) Georgeville Group, Antigonish Highlands. Note sedimentary structures in the centre of the photo indicate that the beds are overturned.

Figure 10.10   Sub-recumbent isoclinal folds (centre left) with basal thrust (visible beneath the fold at the contact between the darker and lighter coloured units), Georgeville Group, Antigonish Highlands.

hornfels facies series. The Fourchu Group was deformed by NE–SW trending folds associated with a steeply dipping cleavage. Clasts of the Fourchu Group containing a predepositional fabric have been found in the overlying Cambrian rocks implying that some deformation occurred prior to deposition of the Cambrian. Farther north in Cape Breton Island, volcanic equivalents of the Fourchu Group are only preserved as outliers. However, correlatives exist in the calc-alkaline plutonic roots to the volcanic edifice (Barr *et al.*, 1988).

Latest Proterozoic rocks in the Antigonish Highlands (Georgeville Group) consist of a bimodal sequence of basalt, basaltic andesite and rhyolite overlain by thick succession of turbidites (Figure 10.9) with arc-derived clasts, and minor basalts (Murphy and Keppie, 1987). The turbidites are predominant in the central Highlands flanked on the north and south by mainly volcanic assemblages. Geochemically, the basalts are within-plate, continental rift tholeiites with some alkalic tendencies, the basaltic andesites are calc-alkalic and the rhyolites have volcanic arc affinities (Murphy *et al.*, 1990). The Georgeville Group is interpreted to represent an intra-arc rift. The palaeogeography of this rift was influenced by contemporaneous motion on the bounding NE–SW faults (Murphy and Keppie, 1987). The continental tholeiitic lavas and abundant feeder dykes and sills coincide with a strong N–S vertical gradient magnetic anomaly and positive gravity anomaly, which may trace the trend of the extensional zone. If this rift zone has preserved its original trend relative to the bounding faults, then a sinistral transtensional origin may be deduced for the intra-arc basin. However, subsequent (Palaeozoic) dextral movements (Murphy *et al.*, 1990) may have rotated the rift in a clockwise sense to the degree that the original orientation of the rift may have been E–W and the original sense of shear would then have been dextral. Polyphase deformation (Figure 10.5) of the Georgeville Group involving thrusts and recumbent east-vergent folds (Figure 10.10) deformed by upright, N–S folds is attributed to dextral transpressional closure of the basin (Murphy *et al.*, 1990). This deformation was accompanied by greenschist–subgreenschist facies metamorphism. In the northern Antigonish Highlands, the F1 folds are cut by the gabbroic Greendale Complex dated at 611 ± 5 Ma by $^{40}Ar/^{39}Ar$ plateau ages on hornblende (Keppie *et al.*, 1989). These appinitic gabbroic-dioritic plutons (Figure 10.11) have a tholeiitic to calc-alkalic chemistry.

In the northern and western Cobequid Highlands, the Late Proterozoic Jeffers Group records a similar stratigraphy and volcanic geochemistry to that of the Georgeville Group (Pe-Piper and Pe-Piper, 1989). In the eastern Cobequid Highlands, the Late Proterozoic Folly River Formation is similar lithologically and geochemically to the upper part of the Georgeville Group (Pe-Piper and Murphy, 1989; Murphy *et al.*, 1990). Deposition of the

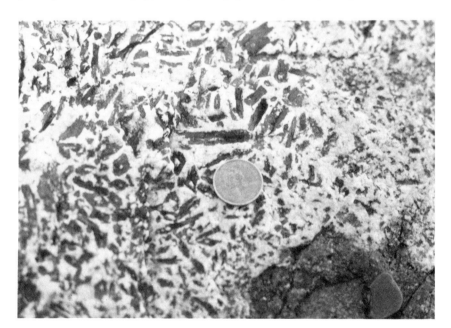

Figure 10.11 Appinitic diorites from the post-tectonic Greendale Complex, Antigonish Highlands composed predominantly of euhedral, prismatic hornblende and plagioclase. Note that hornblende crystals commonly enclose plagioclase.

formation probably post-dates the *c.* 630–610 Ma transtensional deformation in the underlying Gamble Brook Formation. The geochemistry is consistent with continental rifting within a volcanic arc (Pe-Piper and Murphy, 1989). Thus the transtensional deformation may herald the development of basinal conditions within the arc. These Late Proterozoic rocks were deformed by thrusts and isoclinal folds with kinematics indicating dextral transpression (Nance and Murphy, 1990). This deformation took place prior to the intrusion of the Debert River pluton at *c.* 610–605 Ma (U–Pb on zircon; Doig *et al.*, 1989). The above geochronological data indicate that deformation of the Folly River Formation took place soon after deposition and may have partly overlapped mylonitic deformation of the Gamble Brook Formation. The Debert River pluton has magmatic arc affinities.

10.2.4   *Early Palaeozoic overstep sequences*

In southern Cape Breton Island, the Late Proterozoic rocks are overlain by Cambro-Ordovician platformal rocks containing an Acado-Baltic fauna. A major component of these rocks in the Boisdale Hills comprises Middle Cambrian, bimodal, within-plate, continental rift lavas (Murphy *et al.*, 1985). In fault contact with these volcanic rocks is the Boisdale Hills pluton dated at *c.* 535–525 Ma (^{40}Ar/^{39}Ar plateau ages on hornblende; Keppie *et al.*, 1990b), which may be the subvolcanic equivalents of the middle Cambrian volcanic rocks. Plutons of similar age have been dated in the Craignish Hills, North Mountain and western Cape Breton Highlands. Elsewhere in Cape Breton Island, Cambro-Ordovician rocks are absent. However, Precambrian rocks in the Cape Breton Highlands are overlain by Ordovician–Silurian sediments and bimodal volcanics. Unfortunately, fossils have not been recovered from these rocks. However, pebbles in the Lower Carboniferous Horton Group surrounding the Mabou Highlands contain Cambro-Ordovician and Siluro-Devonian fauna typical of the Avalon Composite terrane (Norman, 1935; J. D. Keppie, unpublished data). Current directions and other pebble lithologies indicate derivation from the Mabou Highlands, suggesting the former presence of the overstep sequence across the Avalon composite terrane.

In the Antigonish Highlands, the Cambro-Ordovician platformal units with an Acado-Baltic fauna (Landing *et al.*, 1980) rest unconformably upon deformed Precambrian rocks. Interbedded with the early Cambrian sedimentary rocks are bimodal, within-plate, continental rift volcanic rocks (Murphy *et al.*, 1985). The Cambro-Ordovician rocks are, in turn, unconformably overlain by a Siluro-Devonian succession. Both of these sequences are typical of the Avalonian overstep sequence. In the Cobequid Highlands, only the Siluro-Devonian overstep sequence is preserved.

Palaeomagnetic data indicate that the Avalon composite terrane had affinities with Gondwana and Armorica during Late Proterozoic–Cambrian times rather than with North America (Johnson and Van der Voo, 1985, 1986). However, they had converged by Siluro-Devonian times (Miller and Kent, 1988). This convergence is also shown by the contrasting Cambrian faunal provinces (Acado-Baltic in the Avalon and Laurentian in North America); with the distinct Rhenish-Bohemian fauna that was present only in

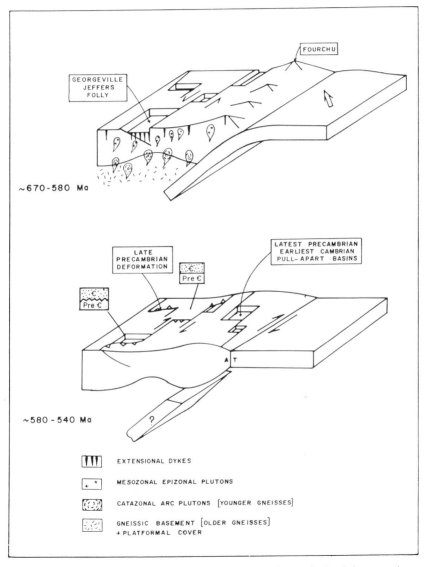

Figure 10.12   Schematic representation of the late Precambrian–early Cambrian tectonic setting in the Avalon composite terrane of Nova Scotia. This is not a palinspastic reconstruction because the actual position of the Fourchu Group relative to the Georgeville and Jeffers groups is unknown.

the Avalon during the Late Silurian–Gedinnian spreading across onto Laurentia during the Siegenian and Emsian (Boucot, 1975).

10.3 Discussion

The tectonic setting of the Avalonian gneisses in Nova Scotia is poorly understood. Some of the gneisses apparently record a Mid- to Late Proterozoic (Grenvillian?) tectono-thermal event or a series of events that occurred prior to the development of the characteristic Late Precambrian volcano-sedimentary sequences. However, it is possible that other gneisses (e.g. Great Village River Gneiss) may represent the metamorphic infrastructure of a late Precambrian magmatic arc. The mid-Riphean (?) miogeoclinal rocks form an extensive platformal succession that probably records the development of a passive margin. They are traditionally interpreted as a 'cover' sequence deposited unconformably on the high grade 'basement'. However, this cannot be demonstrated unequivocally because the present contacts, where exposed, are sheared.

The stratigraphic and petrologic data for the Late Proterozoic rocks of the Avalon composite terrane indicate that they were formed in a rifted ensialic magmatic arc environment in which alternating horsts and grabens were

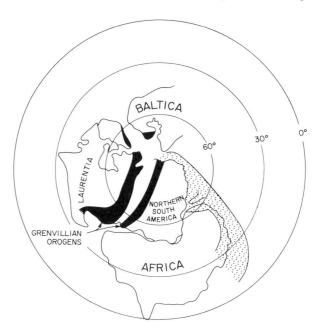

Figure 10.13   Latest Precambrian–earliest Cambrian palinspastic reconstruction of the Circum-Atlantic region (modified after Bond *et al.*, 1984) showing the locations of the Avalonian–Cadomian belt (stipple) and the Grenville orogens (black).

developed (Figure 10.12). Available data indicate that the basins were floored by continental basement. Deformation is associated with the opening and closing of these intra-arc rifts and took place by transcurrent motions. Thus the age of deformation has only local significance and does not yield direct information as to the timing of the Avalonian orogeny because in this type of setting, basin opening and closure may be diachronous. The opening stage was presumably related to oblique convergence. Geochronological data, at present, suggest that the magmatic arc was active from c. 635–600 Ma. The Late Proterozoic magmatic arc evolved into latest Proterozoic–Cambrian tholeiitic-alkaline volcanic rocks. The Cambrian rocks appear to have been deposited in pull-apart basins suggesting a continuation of the transpressional strike-slip regime operative in the Late Proterozoic.

Faunal provinciality and palaeomagnetic data indicate that the Avalon composite terrane had affinities with Gondwana and Armorica during Late Proterozoic–Cambrian times. In most of these regions the Late Proterozoic orogens are built upon Archaean and Early Proterozoic basement. Only in South America are Late Proterozoic orogens built upon a Mid–Late Proterozoic (Grenvillian) basement (Keppie and Dallmeyer, 1989). This continent may therefore represent the provenance of the Avalon composite terrane in Nova Scotia. The synchroneity of Late Proterozoic–Cambrian subduction beneath the Avalon composite terrane and the opening of Iapetus implies that the Avalon was not within the Iapetus Ocean. A possible Late Proterozoic–Cambrian reconstruction is shown in Figure 10.13 where the Avalonian belt is positioned at the northern extremity of a Late Proterozoic supercontinent. If so, this reconstruction suggests that subduction was predominantly southward-directed. The good state of preservation of the Avalonian magmatic arc sequences contrasts with their general elimination in continent-continent collision orogens where they are usually eroded away because they lie on the upper plate. This suggests that termination of subduction was not due to continent-continent collision. The termination of the Avalonian cycle may have occurred in two stages. Firstly, the separation of Baltica from South America during the inception of Iapetus would have required the propagation of a ridge into the Avalonian subduction system. This would fundamentally change the vector of plate motion. If so, the interaction of these two systems may have resulted in termination of subduction under northern mainland Nova Scotia and its replacement by transform faults. Termination of subduction in these areas may have resulted from the relocation of subduction zones into Iapetus as it passed from a passive to an active margin ocean in the late Cambrian.

Acknowledgements

The work undertaken in this paper was funded under the Canada-Nova Scotia Mineral Development Agreement 1985–1990, and is published with the permission of the Director of the

Mineral Resources Division of the Nova Scotia Department of Mines and Energy. J. B. Murphy and J. Dostal acknowledge the support of NSERC (Canada). It also forms a contribution to the International Geological Correlation Programme, Project No. 233: Terranes in Circum-Atlantic Palaeozoic Orogens. We acknowledge the contribution of Sandra Barr, Ken Currie, Howard Donohoe, Becky Jamieson, Georgia Pe-Piper, David Piper, Rob Raeside and Peter Wallace to the understanding of the Avalon terrane in Nova Scotia.

References

Barr, S. M. and Raeside, R. P. (1986) Pre-Carboniferous tectonostratigraphic subdivisions of Cape Breton Island, Nova Scotia. *Mar. Sed. Atlantic Geol.* 22, 252–263.

Barr, S. M. and Raeside, R. P. (1989) Tectonostratigraphic terranes in Cape Breton Island, Nova Scotia: Implications for the configuration of the Northern Appalachian orogen. *Geology* 17, 822–825.

Barr, S. M., O'Reilly, G. A. and O'Beirne, A. M. (1982) Geology and geochemistry of selected granitoid plutons of Cape Breton Island. *N. S. Dep. Mines Energy Pap.* 82–1, 176 pp.

Barr, S. M., Raeside, R. P. and Van Breemen, O. (1987) Grenvillian basement in the northern Cape Breton Highlands, Nova Scotia. *Can. J. Earth Sci.* 24, 992–999.

Barr, S. M., Raeside, R. P., Dunning, G. R. and Jamieson, R. A. (1988) New U–Pb ages from the Cape Breton Highlands and correlations with southern Newfoundland. Lithoprobe East Contribution.

Bond, G. C., Nickerson, P. A. and Kominz, M. A. (1984) Break up of a supercontinent between 625 Ma and 555 Ma: new evidence and implications for continental histories. *Earth Planet. Sci. Lett.* 70, 325–345.

Boucot, A. J. (1975) *Evolution and Extinction Rate Controls.* Developments in Palaeontology and Stratigraphy 1, Elsevier, Amsterdam.

Cullen, M. P. (1984) Geology of the Bass River Complex, Cobequid Highlands, Nova Scotia. M.Sc. thesis, Dalhousie University, Halifax, Nova Scotia, 183 pp.

Currie, K. L. (1987) Relations between metamorphism and magmatism near Cheticamp, Cape Breton Island, Nova Scotia. *Geol. Surv. Can. Pap.* 85–23, 66 pp.

Doig, R., Murphy, J. B. and Nance, R. D. (1989) Preliminary results from U–Pb geochronology, Cobequid Highlands, Nova Scotia. *Geol. Assoc. Canada Abstr. Programs* 14, A126.

Donohoe, H. V. (1976) The Cobequid Mountains Project. *N. S. Dep. Mines Rep.* 76–2, 113–124.

Donohoe, H. V. and Cullen, M. (1983) Deformation, age and regional correlation of the Mt. Thom and Bass River Complexes, Cobequid Highlands, Nova Scotia. *Geol. Soc. Am. Abstr. Programs* 15 (3).

Helmstaedt, H. and Tella, S. (1973) Pre-Carboniferous structural history of S.E. Cape Breton Island, Nova Scotia. *Mar. Sed.* 9, 88–99.

Hoffman, H. (1974) The stromatolite Archaezoan acadiense from the Proterozoic Green Head Group St. John, New Brunswick. *Can. J. Earth Sci.* 11, 1098–1115.

Jamieson, T. S. (1984) Low pressure cordierite-bearing migmatites from Kellys Mountain, Nova Scotia. *Contrib. Mineral. Petrol.* 86, 309–320.

Johnson, R. J. E. and Van der Voo, R. (1985) Middle Cambrian paleomagnetism of the Avalon terrane in Cape Breton Island, Nova Scotia. *Tectonics* 4, 629–651.

Johnson, R. J. E. and Van der Voo, R. (1986) Paleomagnetism of the Late Precambrian Fourchu Group, Cape Breton Island, Nova Scotia. *Can. J. Earth Sci.* 23, 1673–1685.

Keppie, J. D. (1979) Geological, structural and metamorphic maps of Nova Scotia. *N. S. Dep. Mines Energy, Map* 79–1.

Keppie, J. D. (1982) The Minas Geofracture. *Geol. Assoc. Can. Spec. Pap.* 24, 263–280.

Keppie, J. D. (1985) The Appalachian collage. In *The Caledonide Orogen—Scandinavia and Related Areas, Part 2*, ed. Gee, D. G. and Sturt, B. A., John Wiley and Sons, Chichester, 1217–1226.

Keppie, J. D. (1989) Northern Appalachian terranes and their accretionary history. *Geol. Soc. Am. Spec. Pap.* 230, 159–192.

Keppie, J. D. and Dallmeyer, R. D. (1989) Map of Pre-Mesozoic terranes in Phanerozoic orogens. *Int. Geol. Correlation Program*, Project 233, 1:5 000 000.

Keppie, J. D., Dostal, J. and Murphy, J. B. (1979) Petrology of the Late Precambrian Fourchu Group in the Louisburg area, Cape Breton Island. *N. S. Dep. Mines, Pap.* **79–1**.

Keppie, J. D., Nance, R. D., Murphy, J. B. and Dostal, J. (1990b) Northern Appalachians. In *Tectono-thermal Evolution of the West African Orogens and Circum-Atlantic Correlatives*, eds. Dallmeyer, R. D. and Lechorche, J. P., Springer-Verlag, Heidelberg (in press).

Keppie, J. D., Dallmeyer, R. D. and Murphy, J. B. (1990b) Tectonic implications of ^{40}Ar/^{39}Ar hornblende ages from the Late Proterozoic–Cambrian plutons in the Avalon Composite Terrane in Nova Scotia, Canada. *Bull. Geol. Soc. Am.* (in press).

Landing, E., Nowlan, G. S. and Fletcher, T. P. (1980) A microfauna associated with early Cambrian trilobites of the Callavia Zone, northern Antigonish Highlands, Nova Scotia. *Can. J. Earth Sci.* **17**, 400–418.

Loncarevic, B. D., Barr, S. M., Raeside, R. P., Keen, C. E. and Marillier, F. (1989) Northeastern extension and crustal expression of terranes from Cape Breton Island, Nova Scotia using geophysical data. *Can. J. Earth Sci.* **26**, 2255–2267.

MacDonald, A. S. and Smith, P. K. (1980) Geology of Cape North area, northern Cape Breton Island, Nova Scotia. *N. S. Dep. Mines Energy Pap.* **80–1**, 60 pp.

Marillier, F., Keen, C. E., Stockmal, G. S., Quinlan, G., Williams, H., Colman-Sadd, S. P. and O'Brien, S. J. (1989) Crustal structure and surface zonation of the Canadian Appalachians: implications of deep seismic reflection data. *Can. J. Earth Sci.* **26**, 305–321.

Miller, J. D. and Kent, D. V. (1988) Paleomagnetism of the Siluro-Devonian Andreas red beds: evidence for an Early Devonian supercontinent. *Geology* **16**, 195–198.

Murphy, J. B. and Keppie, J. D. (1987) The stratigraphy and depositional environment of the late Precambrian Georgeville Group, Antigonish Highlands, Nova Scotia. *Mar. Sed. Atlantic Geol.* **23**, 49–61.

Murphy, J. B., Cameron, K., Dostal, J., Keppie, J. D. and Hynes, A. J. (1985) Cambrian volcanism in Nova Scotia, Canada. *Can. J. Earth Sci.* **22**, 599–606.

Murphy, J. B., Pe-Piper, G., Nance, R. D. and Turner, D. (1988) A preliminary report on geology of the eastern Cobequid Highlands, Nova Scotia. *Geol. Surv. Can. Pap.* **88–1B**, 99–107.

Murphy, J. B., Keppie, J. D., Dostal, J. and Hynes, A. J. (1990) Georgeville Group: Late Precambrian volcanic arc rift succession in the Avalon terrane of Nova Scotia. In *The Cadomian Orogeny*, eds. D'Lemos, R. S., Strachan, R. A. and Topley, C. G., *Spec. Publ. Geol. Soc. London* **51**,

Nance, R. D., Currie, K. L. and Murphy, J. B. (1990) The Avalon Zone in New Brunswick (Chapter 11, this volume).

Nance, R. D. and Murphy, J. B. (1990) Preliminary kinematic analysis of the Bass River Complex, Cobequid Highlands, Nova Scotia. In *The Cadomian Orogeny*, eds. D'Lemos, R. S., Strachan, R. A. and Topley, C. G., *Spec. Publ. Geol. Soc. London* (in press).

Norman, G. W. H. (1935) Lake Ainslie Map-area, N.S. *Geol. Sur. Can. Mem.* **177**, 103 pp.

Olszewski, W. J. Jr., Gaudette, H. E., Keppie, J. D. and Donahoe, H. V. (1981) Rb–Sr whole rock age of the Kellys Mountain basement complex, Cape Breton Island. *Geol. Soc. Am. Abstr. Programs* **13**, 169.

Pe-Piper, G. and Murphy, J. B. (1989) Petrology of the late Proterozoic Folly River Formation, Cobequid Highlands, Nova Scotia: a continental rift within a volcanic arc environment. *Atlantic Geology* **25**, 143–151.

Pe-Piper, G. and Pe-Piper, D. J. W. (1989) The Late Hadrynian Jeffers Group, Cobequid Highlands, Avalon Zone of Nova Scotia: a back arc volcanic complex. *Bull. Geol. Soc. Am.* **101**, 364–376.

Raeside, R. P. (1987) Geology and metamorphism of the Lime Hill gneissic complex, Inverness County, Cape Breton Island. *Nova Scotia Dept of Mines and Energy Report* 87–5, 193–196.

Sangster, A. L. and Thorpe, R. T. (1988) Marble-hosted zinc occurrences, Cape Breton, Nova Scotia. Abstracts with Programs. *Geol. Soc. Am., Abstr. Programs*, **20** (1).

Stockmal, G. S., Colman-Sadd, S. P., Keen, C. E., Marillier, F., O'Brien, S. J. and Quinlan, G. (1989) Plate tectonic evolution of the Canadian Appalachians: constraints imposed by deep seismic reflection data. *Tectonics* (in press).

Williams, H. and Hatcher, R. D., Jr (1982) Suspect terranes and accretionary history of the Appalachian Orogen. *Geology* **10**, 530–536.

11 The Avalon Zone of New Brunswick

R. D. NANCE, K. L. CURRIE and J. B. MURPHY

11.1 Zone definition

The Avalon Zone of southern New Brunswick forms part of a distinctive tectonostratigraphic belt within the Northern Appalachian orogen that extends discontinuously from offshore eastern Newfoundland to southeastern Massachusetts (Figure 11.1) and is distinguished on the basis of its pre-Silurian evolution (e.g. Williams, 1979). The zone, which constitutes one of the largest Appalachian suspect terranes, is characterized by the presence of volcanic-sedimentary sequences and co-genetic granitoid plutons of late Precambrian (c. 630–550 Ma) age, and by early Palaeozoic platformal successions that contain Acado-Baltic (Atlantic-realm) trilobite fauna (e.g. O'Brien et al., 1983; Rast and Skehan, 1983; Nance, 1986; Keppie et al., 1990). Siluro-Devonian overstep sequences, where present, contain Rhenish-Bohemian fauna.

The zone is separated from the Meguma terrane by the Cobequid-Chedabucto Fault of Nova Scotia and from inboard terranes by a series of major faults that include the Honey Hill-Lake Char-Bloody Bluff system of New England, the Turtle Head-Honeydale system of Maine and New Brunswick, and the Hermitage Bay-Dover system of Newfoundland.

11.2 Zone boundaries

In southern New Brunswick, late Precambrian and Cambro-Ordovician rocks characteristic of the Avalon Zone occupy much of the Caledonia Highlands (Figure 11.2) but are best represented in the vicinity of Saint John where they outcrop south of the Belleisle and Wheaton Brook faults (Figure 11.3). The former records late Palaeozoic dextral motion (Leger and Williams, 1986), while the latter is a south-side-up, dip-slip fault (Thomas and Willis, 1989) stitched by the 367 ± 2 Ma (U–Pb zircon, Bevier, 1988) Mount Douglas granite. However, both faults have probably experienced earlier movement and neither are likely to be terrane boundaries since late Precambrian granites and Silurian rocks containing Rhenish-Bohemian fauna occur to the north of them (Currie, 1988). Keppie et al. (1990) suggest that the northern

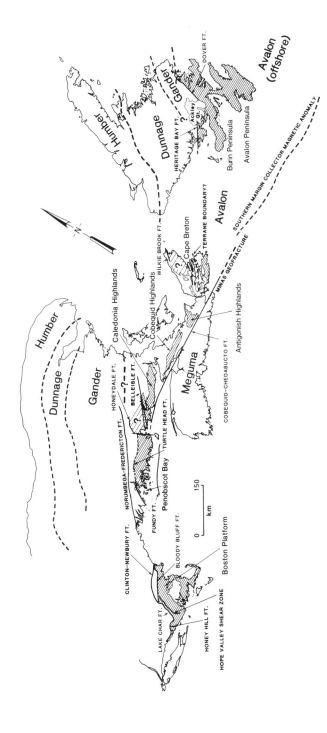

Figure 11.1 Distribution of the Avalon Zone (broad shaded where uncertain) and major terrane boundary faults in the Northern Appalachians (redrawn and adapted from Nance, 1986).

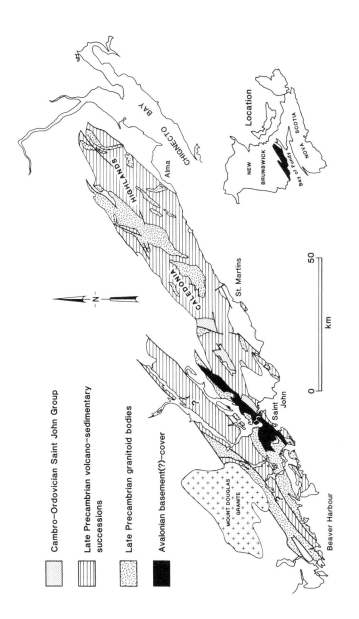

Figure 11.2 Late Precambrian–early Palaeozoic geology of southern New Brunswick (geology simplified after Giles and Ruitenberg, 1977 and Currie, 1988).

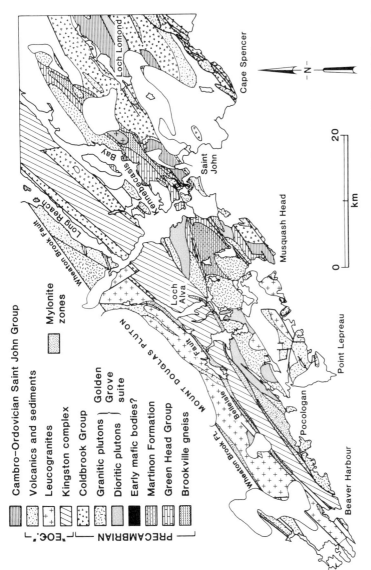

Figure 11.3 Late Precambrian–early Palaeozoic geology of the Saint John region of southern New Brunswick (simplified after Currie, 1988). See Figure 11.2 for location.

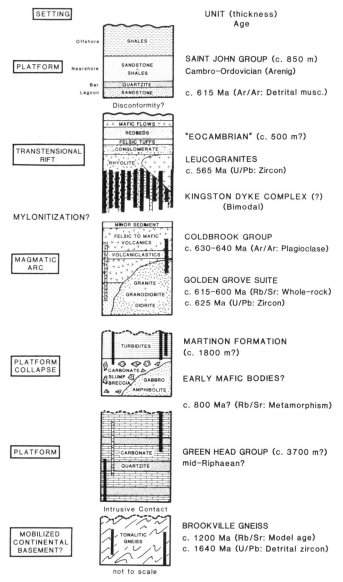

Figure 11.4 Interpretive, Precambrian to early Palaeozoic tectono-stratigraphic column for the Avalon Zone in southern New Brunswick (redrawn and adapted from Nance, 1987).

boundary of the Avalon Zone coincides with the Honeydale Fault (Figure 11.1) which has been interpreted as a northwest-vergent thrust (Ruitenberg and Ludman, 1978). The southern margin of the zone lies beneath the Bay of Fundy where it takes the form of a south-dipping, late Palaeozoic listric thrust fault which separates the Avalon Zone from the Meguma terrane of southern Nova Scotia (Figure 11.1). The boundary was reactivated as a normal fault during the Triassic opening of the Bay of Fundy.

11.3 Unit descriptions

The names, lithologies, and proposed tectonic settings for major units in southern New Brunswick have been described by Ruitenberg *et al.* (1979), Currie (1986, 1988), and Nance (1987, 1990), and are summarized in Figure 11.4 in the form of an interpretive tectonostratigraphic column. The succession, which is generally thought to have been built upon continental basement of Mid-Proterozoic age represented by the Brookville gneiss, includes mid-Riphean(?) shelf carbonates and quartzites of the Green Head Group; Late Proterozoic carbonate conglomerates, greywackes and basalts of the Martinon Formation; minor Late Proterozoic amphibolites; latest Proterozoic calc-alkaline granitoid plutons and arc-related volcanics and volcaniclastics of the Coldbrook Group; a bimodal dyke swarm (Kingston

Figure 11.5 Brookville paragneiss with mildly discordant enclaves of light-coloured biotite granite orthogneiss.

complex) that is correlated with bimodal volcanics, plutons and redbeds of 'Eocambrian' age; and Cambro-Ordovician sandstones and shales of the Saint John Group.

Younger units are beyond the scope of this chapter but include fault-bound basalts, tuffs, sandstones and siltstones of Silurian age; Late Silurian to mid-Devonian granitoid bodies; and Late Devonian, Carboniferous and Triassic redbeds.

11.3.1   Brookville gneiss

Possibly the oldest unit in southern New Brunswick, the tonalitic Brookville gneiss (Figure 11.5) is thought to represent ductilely mobilized and partially melted basement to the Avalon Zone (Currie et al., 1981). This quartz-plagioclase-hornblende ± biotite paragneiss has been metamorphosed to the upper amphibolite facies (Wardle, 1978) and contains relict mafic dykes not represented in younger units. Enclaves of biotite granite orthogneiss were apparently developed through in situ anatexis (Currie et al., 1981) while tourmaline pegmatites may be younger.

Existing radiometric data do not define a satisfactory age for the Brookville gneiss. Olszewski and Gaudette (1982) analysed zircons of detrital morphology and interpreted the results in terms of an original age of 1641 ± 60 Ma with Pb-loss events at 783 ± 40 and 369 ± 45 Ma. A single metamorphic zircon gave an essentially concordant age of 814 Ma. Metamorphic zircons from a quartz diorite gneiss yielded upper and lower intercept ages of 827 ± 40 Ma and 333 ± 40 Ma based on two analyses. Rb/Sr whole-rock data for both gneisses gave an age of 771 ± 55 Ma. Collectively, these data suggest high grade metamorphism of the Brookville gneiss took place at about 800 ± 30 Ma. On the basis of initial strontium ratios, Olszewski and Gaudette (1982) suggested that the original age of the Brookville gneiss is unlikely to exceed 1200 Ma. If so, the age of the protolith would be consistent with Grenvillian (c. 1100 Ma) basement (Currie et al., 1981; Currie, 1986; Nance 1987).

Contact relations with the adjacent Green Head Group are those of a ductile shear zone, with dyke-like projections of gneiss invading the Green Head Group and ductilely mobilized Green Head marbles and locally sillimanite-bearing metaclastics incorporated in the gneiss (Currie et al., 1981). The steep thermal gradient adjacent to the gneiss implied by the predominantly greenschist facies Green Head Group, coupled with their contrast in composition, the steep attitude of the contact between them, and the multidirectional, steeply plunging and polyphase nature of the gneissic folding, has been taken to indicate emplacement of the gneiss through hot gneissic diapirism (Wardle, 1978; Currie et al., 1981). The development of these ductile contact relations is thought to have accompanied regional metamorphism at c. 800 Ma (Currie, 1986; Nance, 1987) at which time the basement is interpreted to have been mobilized under amphibolite facies conditions and diapirically emplaced into the overlying Green Head cover. However, hornblendes from the Brookville gneiss have recently yielded a ^{40}Ar/^{39}Ar plateau age of 548 ± 5 Ma (Dallmeyer and Nance, 1989) suggesting a much younger tectonothermal history than was previously supposed. Preliminary U–Pb ages of c. 610 Ma from orthogneiss zircons (M. L. Bevier, pers. commun., 1989) also suggest a much younger protolith age and bring into question the interpretation of the Brookville Gneiss as basement to the Green Head Group.

11.3.2   Green Head Group

The Green Head Group comprises a thick sequence of colour-banded marbles, massive dolomites, orthoquartzites, and minor meta-siltstones and pelitic schists. Wardle (1978) informally subdivided this carbonate-clastic sequence into three formations; namely a basal, predominantly clastic and locally cross-bedded (Figure 11.6) Lily Lake Formation; a largely carbonate Drury Cove Formation; and an uppermost, heterogeneous Narrows Formation comprising minor conglomerates, calcareous pelites and carbonates which locally contain the mid-Riphean(?) (Hofmann, 1974) stromatolite *Archaeozoon acadiense* (Figure 11.7). Wardle (1978) interpreted this stratigraphy to reflect depositional environments that evolved from a stable carbonate shelf to basinal slope conditions. However, given the polyphase and heterogeneous nature of the group's deformation (Nance, 1982), the details of this stratigraphy are open to question.

Figure 11.6   Cross-bedded orthoquartzite of the Green Head Group.

Figure 11.7   Stromatolite *Archaeozoon acadiense* (Hofmann, 1974) in the Green Head Group.

With the exception of contact zones bordering the Brookville gneiss, where sillimanite-bearing pelitic assemblages are locally developed, the bulk of the Green Head Group lies within the greenschist facies. However, some stable assemblages reflect the retrogression of higher grade phases from an earlier regional metamorphism while others define retrograded contact metamorphic aureoles around late Precambrian plutons (Wardle, 1978). Several generations of cross-cutting mafic dykes, some of which may be correlative with the Kingston complex, preserve low-grade metamorphic assemblages. These dykes cut the earliest tectonic fabric of the Green Head Group, which Wardle (1978) considered to be of Precambrian age, but were deformed during a later deformation. Rare felsic dykes also show low-grade metamorphic assemblages and are compositionally similar to acid volcanics in the overlying Coldbrook Group.

11.3.3   *Martinon Formation*

The Martinon Formation comprises a homogeneous (*c*. 1800 m?) clastic assemblage of massive grey sandstones, greywackes, siltstones and basalt, floored by a conglomerate of Green Head limestone and occasional black chert (Figure 11.8). The formation is traditionally included in the Green Head Group but differs from the underlying carbonate-clastic succession in lithology, sedimentology and structural style and was considered a separate unit by Currie (1984). The time-gap represented by the basal unconformity is of unknown duration. There is no recorded structural discordance across it and both successions are cut by mafic dykes of probable late Precambrian

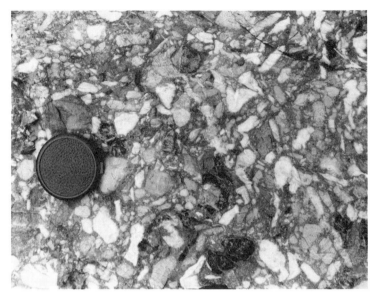

Figure 11.8   Basal Green Head carbonate-chert breccia of the Martinon Formation.

age (Wardle, 1978). Hence the age of the Martinon Formation remains uncertain but it may be broadly equivalent to lithologically similar portions of the Burin Group (*c.* 760 Ma; Krogh *et al.*, 1988) of Newfoundland. The

Figure 11.9   Early amphibolite gneiss with plagioclase pegmatite sweats and xenoliths of Brookville paragneiss (right of coin).

formation is cut by late Precambrian (c. 615 Ma) granitoid plutons and overlain with approximate conformity by the Coldbrook Group. The stratigraphy of the formation, in which basal slump breccias of Green Head carbonate are overlain by an upward-deepening succession of turbiditic sandstones and siltstones, appears to record the abrupt subsidence, slumping and structural disruption of the Green Head continental platform (Nance, 1987).

11.3.4   Early amphibolites?

Carbonates of the Green Head Group are locally intruded by small, occasionally serpentine-rich amphibolite bodies (Figure 11.9). These 'Early Gabbros' of Wardle (1978) appear to show the same metamorphic history as the Green Head carbonates and may be locally responsible for contact metamorphism that predates the earliest tectonic fabric of their host. Nance (1987) suggested their emplacement may be broadly contemporaneous with the platform collapse recorded in the Martinon Formation and tentatively interpreted both to be early manifestations of the c. 800 Ma tectono-thermal event recorded in the Brookville gneiss. However, no contact relations between the Martinon Formation and the early mafic bodies are preserved and the geochemical affinities of the amphibolites are unknown. Given their present uncertainty in age, therefore, it is equally possible that both units immediately predate the late Precambrian Coldbrook Group.

11.3.5   Late Precambrian granitoid plutons (Golden Grove suite)

Widespread emplacement of late Precambrian granitoid plutons ('Golden Grove suite' of Hayes and Howell, 1937) followed development of the earliest tectonic fabric in the Green Head Group (Wardle, 1978) and is thought to have broadly accompanied the volcanism of the Coldbrook Group. The suite comprises a diverse assemblage of biotite/hornblende-bearing rocks that range from granite through granodiorite to tonalite and diorite and show I-type, calc-alkaline chemistry related to subduction (Dickson, 1985; Barr, 1987). Evidence of acid-basic magma comingling (Figure 11.10) is ubiquitous.

The plutons intrude all of the previously described units and are preserved as pebbles at the base of the Cambro-Ordovician succession (Currie, 1984) although the actual unconformity is not exposed. Pluton contacts with the Green Head Group and Martinon Formation are sharp, and retrograded andalusite-bearing contact aureoles overprint the earliest tectonic fabric of the host rock (Wardle, 1978). Contacts with the Brookville gneiss are gradational, hybridized and metasomatic, the pluton margins tending to be potassic and megacrystic (Currie et al., 1981). In the Caledonia Highlands (Figure 11.2), the plutons intrude volcanics of the Coldbrook Group although both suites are considered to be roughly contemporary and possibly co-

Figure 11.10   Acid-basic magma comingling in the Golden Grove suite. Note chilled margins
(dark) of mafic pillows against host grandiorite.

genetic on the basis of available age data and their subduction-related
geochemistries (Barr, 1987).

The Golden Grove suite has been dated by a U–Pb zircon age of
625 ± 15 Ma (Watters, 1987) from a granite at Cape Spencer (Figure 11.3) and
a Rb–Sr age of 615 ± 37 Ma (Olszewski and Gaudette in Poole, 1980) from
the Musquash area. Two ages of 598 ± 18 Ma (whole-rock Rb–Sr) and
598 ± 27 Ma (K–Ar, hornblende) from granitic and dioritic plutons northwest
of Alma (Figure 11.2) in the eastern Caledonia Highlands (Barr, 1987)
probably date the same magmatic episode.

11.3.6   *Coldbrook Group*

The Coldbrook Group comprises a thick succession of volcanics, volcaniclastics and volcanogenic sediments (Giles and Ruitenberg, 1977) whose
stratigraphy has yet to be established in detail due to problems of correlation.
Dyke relations and inclusions suggest that the group rests on the Brookville

Figure 11.11   Dacitic volcanic breccia (lahar?) in the Coldbrook Group.

gneiss and Green Head Group although actual contacts are everywhere faulted. The group is unconformably overlain by 'Eocambrian' and Lower Cambrian strata (Hayes and Howell, 1937; Currie, 1988) and is likely to be co-genetic with respect to the late Precambrian Golden Grove suite. Available radiometric age determinations are, as yet, unsatisfactory but the best estimate appears to be the $^{40}Ar/^{39}Ar$ plagioclase age range of 640–630 Ma obtained by Stukas (1977). The Coldbrook Group is penetratively deformed and metamorphosed to the prehnite-pumpellyite facies.

In the Saint John region, the Coldbrook Group consists of basal tholeiitic basalt flows and tuffs, overlain by a thick pile of andesite to dacite volcanic breccias and pyroclastics (Figure 11.11), and capped by intermediate to acid tuff and volcanogenic conglomerates and cherty siltstones (Currie and Eby, 1990). A similar tholeiitic base has been demonstrated in the eastern Caledonia Highlands (Barr and White, 1988), where it is intruded by correlatives of the Golden Grove suite. Geochemical studies in both regions clearly demonstrate a subduction-related origin in which arc magmatism was derived from a distinctive U-depleted mantle source and emplaced into continental or transitional crust (Currie and Eby, 1990). Older models favouring an intracratonic rift as the tectonic setting for the Coldbrook Group (Giles and Ruitenberg, 1977; Ruitenberg et al., 1979) were based on a stratigraphy that failed to distinguish between Coldbrook and 'Eocambrian' volcanic rocks.

Figure 11.13   Deformed granite of the Golden Grove suite showing dextral sense of shear (c's) in the Pocologan mylonite zone.

11.3.7   Kingston dyke complex

The Kingston complex comprises a bimodal dyke swarm (Figure 11.12) which defines a linear zone up to 10 km across strike that extends over 150 km from Carboniferous cover northeast of Saint John to the US/Canadian border (Currie, 1984). The complex is bordered on both sides by major mylonite or brittle fault zones (Figure 11.3). However, outlying dykes and inclusions within the complex suggest it originally intruded the Green Head Group and Golden Grove suite as well as correlative bimodal volcanics of the 'Eocambrian' succession. The dykes trend NE–SW in southwestern portions of the belt (Beaver Harbour swarm of Rast, 1979) but trend north–south in the northeastern part (Currie, 1984). Alternating tholeiitic and rhyolitic-dacitic dykes with widths of 1–20 m are typical, but parts of the complex show a 70% mafic to felsic ratio and individual dykes reach widths of 100 m.

The age of the dyke swarm is uncertain although older parts of the complex are pervasively mylonitized while the youngest dykes are undeformed. However, the dykes do not cut the 'Eocambrian' and Lower Palaeozoic

Figure 11.13   Deformed granite of the Golden Grove suite showing dextral sense of shear (c's) in the Pocologan mylonite zone.

successions and are present as inclusions in a leucogranite body of presumed 'Eocambrian' age (Lingley pluton) at the southwest end of Long Reach (Figure 11.3). A Late Precambrian emplacement age (Rast, 1979) therefore seems likely. However, a preliminary U–Pb zircon age of c. 435 Ma (R. Doig, pers. commun., 1989) from rhyolite dykes northwest of Loch Alva (Figure 11.3) suggests that part of the complex is of lowermost Silurian age.

Mafic members of the dyke complex are mainly rift-related continental tholeiites (Dickson, 1985) while acid members exhibit distinctive Th-rich compositions. The oblique trends of the dykes relative to the axis of the Kingston complex suggest emplacement accompanied sinistral shear (Leger and Williams, 1988). Their emplacement has been interpreted to have coincided with latest Precambrian sinistral transtension (Currie, 1988) which is considered to have produced the distinctive 'Eocambrian' bimodal magmatic suite.

11.3.8   *Late Precambrian*(?) *mylonite zones*

West of Loch Alva (Figure 11.3), both margins of the Kingston complex are affected by intense ductile shear (Rast and Dickson, 1982; Currie, 1988) with well developed kinematic indicators (Figure 11.13) of dextral shear sense (Leger and Williams, 1986). A parallel zone of mylonitization follows the Wheaton Brook Fault southwest of the Mount Douglas pluton. All three zones are followed by younger brittle faults that are, at least locally, of Carboniferous age. However, the age of ductile deformation is less certain. An Acadian age for ductile shear on the southern (Pocologan) mylonite zone (Leger and Williams, 1986) is supported by a new ^{40}Ar/^{39}Ar plateau age of 411 ± 2 Ma (Dallmeyer and Nance, 1989) obtained from hornblendes separated from a mylonitic mafic dyke northwest of Pocologan. However, field relationships suggest earlier ductile motion. The dextral fabric of the central (Seven Mile Lake) mylonite zone is locally cross-cut by distinctive and essentially undeformed dykes of the Kingston complex and, at Beaver Harbour (Figure 11.3), is overlain by mildly deformed, fossiliferous Cambrian strata along a contact that does not appear to record significant movement. Mylonite occurs in rotated inclusions within the Lingley pluton which grades into 'Eocambrian' volcanic strata underlying the Cambro-Ordovician Saint John Group. Quartz-mylonite clasts and deformed granitoid and volcanic pebbles are preserved in the Tommotian Ratcliffe Brook Formation of the Saint John Group and large detrital muscovites from the same formation have yielded a ^{40}Ar/^{39}Ar plateau age of 615 ± 2 Ma (Dallmeyer and Nance, 1989). These muscovites probably date a phase of ductile shear because only within mylonite zones is muscovite widely developed in the late Precambrian granitoid bodies. Hence, the mylonite zones probably record polyphase ductile shear of latest Precambrian and latest Silurian age. Late Precambrian mylonitization, if present, may record the closure of the rift now represented

Figure 11.14   Pillow breccia of amygdaloidal basalt (hyaloclastite) in the 'Eocambrian' succession.

Figure 11.15   'Eocambrian' ignimbrite with flattened fragments of darker pumice (fiamme).

by parts of the Kingston complex under conditions of dextral shear (Nance, 1987; Currie, 1988).

11.3.9   *'Eocambrian' succession*

The 'Eocambrian' succession recognized by Currie (1984) comprises a culminating sequence of amygdaloidal basalt (Figure 11.14), rhyolite porphyry, and tuff (Figure 11.15), and red feldspathic sandstone and shale which unconformably overlie the Coldbrook Group and pass upwards into the basal Cambrian Ratcliffe Brook Formation. The succession is likely to include the bimodal, continental tholeiites and felsic pyroclastics discussed by Greenough *et al.* (1985) and is contemporaneous with distinctive bimodal, high level plutons, with which it locally shows gradational contacts. Granitic members of this plutonic suite have been dated at 565 ± 8 Ma (U–Pb, zircon; Currie, 1988) and exhibit a distinctive Th-rich composition similar to the rhyolites of the Kingston complex. Mafic members range from diorite to metagabbro and are enriched in platinum group elements. This volcanic-plutonic suite is at least as voluminous as the Coldbrook-Golden Grove association in both the Saint John region and the eastern Caledonia Highlands (Barr and White, 1988) and has been interpreted as a rift-related package which marks the transition from older subduction-related magmatism to the platformal conditions recorded in the overlying Saint John Group (Nance, 1987).

11.3.10   *Saint John Group*

The Cambrian to Lower Ordovician (Arenig) shelf quartzites, sandstones and shales of the Saint John Group (Hayes and Howell, 1937; Tanoli and Pickerill, 1988) contain Acado-Baltic fauna and represent an overstep sequence to the Precambrian evolution of the New Brunswick Avalon Zone. The basal Cambrian Ratcliffe Brook Formation comprises purple to grey conglomerates, sandstones, siltstones and shales which disconformably(?) overlie redbeds of the 'Eocambrian' succession (Figure 5a, Tanoli and Pickerill, 1988). Trace fossils suggest alluvial, lagoonal and shallow marine environments. The formation is overlain by white orthoquartzites and quartz-pebble conglomerates of the Glen Falls Formation which are interpreted as the remnants of a barrier island system. Middle Cambrian strata include the *Paradoxides-* and *Agnostus*-bearing, grey sandstone-shale, marine shelf sequence of the Hanford Brook and Forest Hills Formations of Tanoli and Pickerill (1988). Sandstones and siltstones of the Hanford Brook Formation contain abundant phosphate nodules and glauconites and were

Figure 11.16   Fine-grained, cross-laminated sandstones of the Middle to Upper Cambrian King Square Formation (Saint John Group).

deposited on a starved, open marine shelf. Massive mudstones of the Forest Hills Formation represent a mud-dominated shelf sequence. Middle Cambrian strata at Beaver Harbour (Figure 11.3) contain within-plate basaltic flows and breccias (Greenough *et al.*, 1985). The upper Middle Cambrian to Lower Ordovician succession includes the King Square (Figure 11.16), Silver Falls and Reversing Falls formations of Tanoli and Pickerill (1988) and consists of basal, wave- and storm-dominated grey shales and sandstones overlain by a deep marine shelf sequence of monotonous black, calcareous shales bearing *Olenus*, *Dictyonema* and *Tetragraptus*.

11.4   Regional correlations

Possible correlations between the Avalon Zone of New Brunswick and that of Nova Scotia and Newfoundland (Figure 11.17) have been reviewed by O'Brien *et al.* (1983), Rast and Skehan (1983), Nance (1986) and Keppie *et al.* (1990). Gneissic complexes analogous to the Brookville gneiss include the Great Village River gneiss and Mount Thom complex of the Cobequid Highlands of Nova Scotia (Figure 11.1) and the Kellys Mountain complex of Cape Breton Island. Rb–Sr errorchrons of 934 ± 32 Ma from Mount Thom (Gaudette *et al.*, 1984) and 701 ± 66 Ma from Kellys Mountain (Olszewski *et al.*, 1981) may provide minimum ages for the metamorphism of these complexes. Rocks of Grenvillian age (*c.* 1040 Ma, U–Pb) occur in northwestern Cape Breton (Barr *et al.*, 1987) but their relationship to the adjacent Avalon Zone is not clear. No basement to the late Precambrian sequences is known in Newfoundland.

Platformal carbonate-quartzite assemblages like that of the Green Head Group are preserved in the George River Group of Cape Breton Island and in the quartzitic Gamble Brook Formation of the Cobequid Highlands. They are generally thought to represent 'cover' successions to the 'basement' gneissic complexes although their original contact relations have not been observed and recent age determinations do not support this. The only indication of the age of the 'cover' successions is the presence of mid-Riphean(?) stromatolites (Hofmann, 1974) in the Green Head Group.

The 762 ± 3 Ma (U–Pb; Krogh *et al.*, 1988) Burin Group of Newfoundland consists mainly of mafic pillow lavas with early alkalic and later oceanic tholeiitic affinities, and minor stromatolitic marble conglomerate and siltstone (Strong *et al.*, 1978). The group, which has been interpreted to have formed in an aborted rift, is broadly contemporary with high temperature-low pressure metamorphic ages in New Brunswick and Nova Scotia (e.g. Keppie *et al.*, 1990), and may be broadly correlative with the Martinon Formation.

Late Precambrian volcanic-sedimentary successions and correlative plutons are characteristic of the Avalon Zone but tend to belong to two temporally, compositionally and tectonically distinct episodes. Earlier vol-

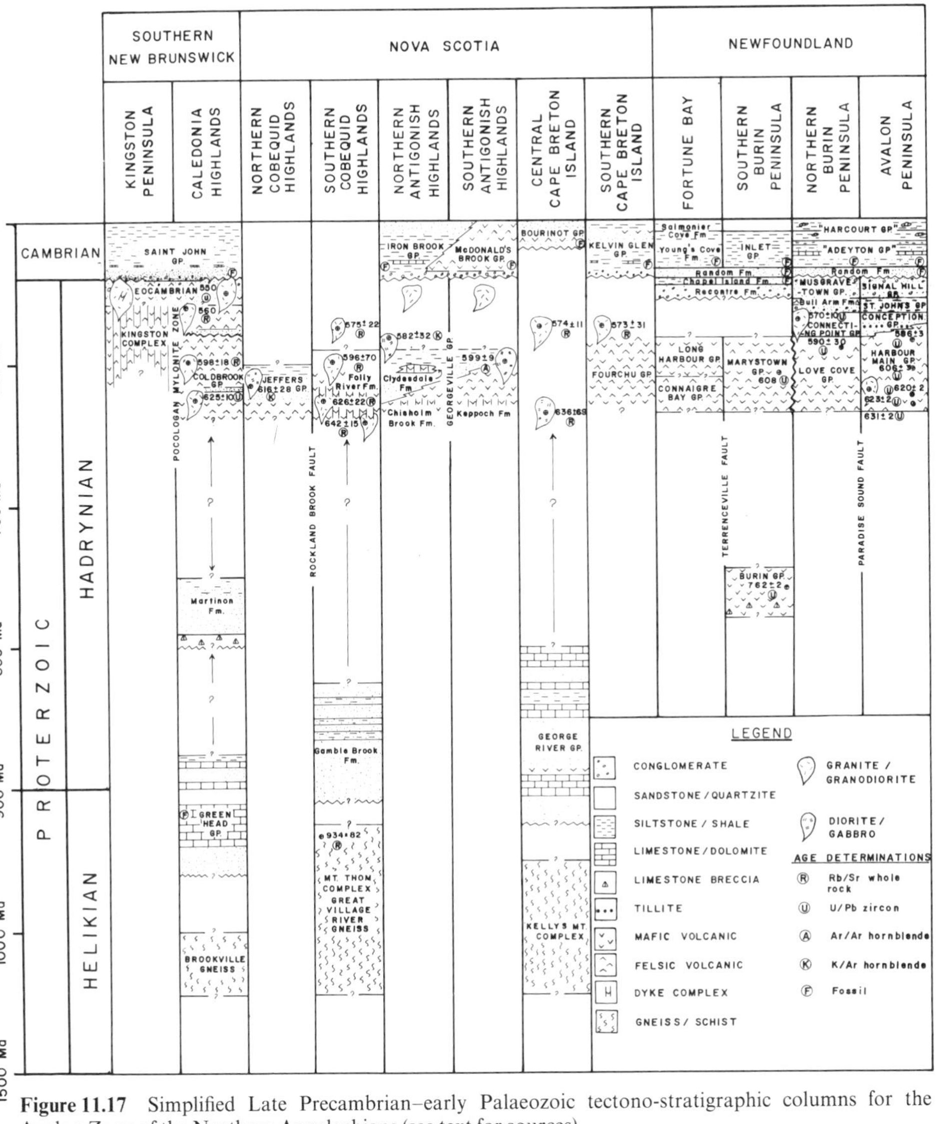

Figure 11.17 Simplified Late Precambrian–early Palaeozoic tectono-stratigraphic columns for the Avalon Zone of the Northern Appalachians (see text for sources).

canic successions such as the Coldbrook Group of New Brunswick, the Keppoch Formation and Fourchu Group of Nova Scotia, and the Connaigre Bay, Long Harbour, Marystown, Love Cove and Harbour Main Groups of Newfoundland, are calc-alkaline and widely interpreted as the products of one or more ensialic volcanic arcs. These successions tend to contain little sedimentary material and may be largely terrestrial. In Nova Scotia, however, the Jeffers Group and Folly River Formation of the Cobequid Highlands and the Georgeville Group of Antigonish Highlands (Figure 11.1) are associated with turbidites and are inferred to have formed in a volcanic arc rift (Murphy *et al.*, 1990). In Newfoundland, volcanics and correlative plutons span the interval 630–585 Ma but were largely emplaced prior to 600 Ma (Krogh *et al.*, 1988) and follow an evolutionary path similar to that of the Coldbrook Group in which initial mildly tholeiitic volcanism is succeeded by voluminous calc-alkaline activity (O'Brien *et al.*, 1988). The succession is followed conformably by synorogenic flysch of the Conception and Connecting Point Groups which may also have formed in volcanic arc rifts. A distinctive tillite unit occurs near the base of the Conception Group.

Younger volcanic successions such as the Bull Arm volcanics of Newfoundland (O'Brien *et al.*, 1988), the Main à Dieu belt of the Fourchu Group in Cape Breton Island (Barr *et al.*, 1988), and the 'Eocambrian' succession of southern New Brunswick (Currie, 1988), are distinctly bimodal and likely to be the products of rifting. Mafic members comprise basalt and gabbro of alkaline affinities, while acid members are high-silica rocks and locally peralkaline. The volcanics are interbedded with and succeeded by red bed sequences such as the Musgravetown and Signal Hill groups of Newfoundland which record terrestrial sedimentation in rift or strike-slip basins and are overlain with slight disconformity by Lower Cambrian strata. Available U–Pb dating suggests a limited age range of 565–550 Ma for this igneous activity although Rb–Sr estimates give ages as low as 530 Ma.

Cambro-Ordovician shallow-marine platformal successions such as the Saint John Group include the Iron Brook and McDonald's Brook groups of the Antigonish Highlands, the Bourinot and Kelvin Glen groups of Cape Breton Island, and the Young's Cove Formation and the Inlet, Adeyton, and Harcourt groups of Newfoundland. These form discontinuous overlap sequences to the zone's late Precambrian lithostratigraphy and contain the distinctive Acado-Baltic fauna. Typically floored by Lower Cambrian (Tommotian) orthoquartzites such as the Glen Falls and Random formations, these sequences are dominated by shales, sandstones, limestones and local oolitic ironstones and have been interpreted as the product of deposition in pull-apart basins (Keppie and Murphy, 1988). Minor but widespread bimodal volcanism of Lower to Middle Cambrian age shows within-plate continental alkalic to tholeiitic affinities (Murphy *et al.*, 1985; Greenough and Papezik, 1986) and may record local dextral transtension persisting from latest Precambrian time.

Acknowledgements

In preparing this chapter the authors are indebted to all those individuals who have contributed to our present understanding of the geology of southern New Brunswick. The contributions of S. Barr, M. L. Bevier, R. Doig, R. Grant, S. McCutcheon, N. Rast, A. Ruitenberg and P. Williams are acknowledged in particular.

References

Barr, S. M. (1987) Field relations, petrology and age of plutonic and associated metavolcanic and metasedimentary rocks, Fundy National Park area, New Brunswick. In *Current Research, Part A, Geol. Surv. Canada Pap.* **87-1A**, 263–280.

Barr, S. M. and White, C. E. (1988) Field relations, petrology, and age of the northeastern Point Wolfe River pluton and associated metavolcanic and metasedimentary rocks, eastern Cobequid Highlands, New Brunswick. In *Current Research, Part A, Geol. Surv. Canada Pap.* **88-1A**, 55–67.

Barr, S. M., Raeside, R. P. and Van Breeman, O. (1987) Grenvillian basement in the northern Cape Breton Highlands, Nova Scotia. *Can. J. Earth Sci.* **24**, 992–997.

Barr, S. M., MacDonald, A. S., White, C. E. and Van Wagoner, N. A. (1988) The Fourchu Group and associated plutonic rocks of southeastern Cape Breton Island. *N. S. Dep. Mines Energy Rep.* **88-3**, 185–188.

Bevier, M. L. (1988) U-Pb geochronologic studies of igneous rocks in New Brunswick. *N. B. Dep. Natural Resour. Inf. Circ.* **88-2**, 134–140.

Currie, K. L. (1984) A reconsideration of some geological relations near Saint John, New Brunswick. In *Current Research, Part A, Geol. Surv. Canada Pap.* **84-1A**, 193–201.

Currie, K. L. (1986) The stratigraphy and structure of the Avalonian terrane around Saint John, New Brunswick. *Mar. Sed. Atlantic Geol.* **22**, 278–295.

Currie, K. L. (1988) The western end of the Avalon zone in southern New Brunswick. *Mar. Sed. Atlantic Geol.* **24**, 339–352.

Currie, K. L. and Eby, G. N. (1990) Geology and geochemistry of the late Precambrian Coldbrook Group near Saint John, New Brunswick. *Can. J. Earth Sci.* **27**, (in press).

Currie, K. L., Nance, R. D., Pajari, G. E., Jr. and Pickerill, R. K. (1981) Some aspects of the pre-Carboniferous geology of Saint John, New Brunswick. In *Current Research, Part A, Geol. Surv. Canada Pap.* **81-1A**, 23–30.

Dallmeyer, R. D. and Nance, R. D. (1989) $^{40}Ar/^{39}Ar$ mineral age record of polyphase tectono-thermal activity in the Avalon terrane of southern New Brunswick. *Geol. Assoc. Can. Mineral. Assoc. Can., Program Abstr.* **14**, A126.

Dickson, W. L. (1985) *Geology, geochemistry and petrology of the Precambrian and Carboniferous rocks between Saint John and Beaver Harbour, southern New Brunswick.* Ph.D. thesis, University of New Brunswick, Fredericton.

Gaudette, H. E., Olszewski, W. J., Jr. and Donohoe, H. V. (1984) Rb/Sr isochrons of Precambrian age from plutonic rocks in the Cobequid Highlands, Nova Scotia. In *N. S. Dep. Mines Energy Rep.* **84-1A**, 285–292.

Giles, P. S. and Ruitenberg, A. A. (1977) Stratigraphy, palaeogeography and tectonic setting of the Coldbrook Group in the Caledonia Highlands of southern New Brunswick. *Can. J. Earth Sci.* **14**, 1263–1275.

Greenough, J. D. and Papezik, V. S. (1986) Acado-Baltic volcanism in eastern North America and western Europe: Implications for Cambrian tectonism. *Mar. Sed. Atlantic Geol.* **22**, 240–251.

Greenough, J. D., McCutcheon, S. R. and Papezik, V. S. (1985) Petrology and geochemistry of Cambrian volcanic rocks from the Avalon Zone in New Brunswick. *Can. J. Earth Sci.* **22**, 881–892.

Hayes, A. O. and Howell, B. G. (1937) *Geology of Saint John, New Brunswick.* Geological Society of America Special Paper **5**.

Hofmann, H. J. (1974) The stromatolite *Archaeozoon acadiense* from the Proterozoic Greenhead Group of Saint John, New Brunswick. *Can. J. Earth Sci.* **11**, 1098–1115.

Keppie, J. D. and Murphy, J. B. (1988) Anatomy of a telescoped pull-apart basin: an example from the Cambro-Ordovician of the Antigonish Highlands. *Mar. Sed. Atlantic Geol.* **24**, 123–138.

Keppie, J. D., Nance, R. D., Murphy, J. B. and Dostal, J. (1990) Northern Appalachians: Avalon and Meguma terranes. In *Tectonothermal Evolution of the West African Orogens and Circum-Atlantic Correlatives*, eds. Dallmeyer, R. D. and Lécorché, J. P., Springer-Verlag, Heidelberg, (in press).

Krogh, T. E., Strong, D. F., O'Brien, S. J. and Papezik, V. S. (1988) Precise U-Pb zircon dates from the Avalon Terrane in Newfoundland. *Can. J. Earth Sci.* **25**, 442–453.

Leger, A. and Williams, P. F. (1986) Transcurrent faulting history of southern New Brunswick. In *Current Research, Part B, Geol. Surv. Can. Pap.* **86-1B**, 111–120.

Leger, A. and Williams, P. F. (1988) Comment on 'Model for the Precambrian evolution of the Avalon terrane, southern New Brunswick, Canada'. *Geology* **16**, 475–476.

Murphy, J. B., Cameron, K., Dostal, J., Keppie, J. D. and Hynes, A. J. (1985) Cambrian volcanism in Nova Scotia, Canada. *Can. J. Earth Sci.* **22**, 599–606.

Murphy, J. B., Keppie, J. D., Nance, R. D. and Dostal, J. (1989) The Avalon composite terrane in Nova Scotia. (This volume).

Nance, R. D. (1982) Structural reconnaissance of the Green Head Group, Saint John, New Brunswick. In *Current Research, Part A, Geol. Surv. Canada Pap.* **82-1A**, 37–43.

Nance, R. D. (1986) Precambrian evolution of the Avalon terrane in the Northern Appalachians: A review. *Mar. Sed. Atlantic Geol.* **22**, 214–238.

Nance, R. D. (1987) Model for the Precambrian evolution of the Avalon terrane in southern New Brunswick, Canada. *Geology* **15**, 753–756.

Nance, R. D. (1990) Late Precambrian–early Palaeozoic evolution of part of the Avalon terrane in southern New Brunswick, Canada. In *The Cadomian Orogeny*, eds. D'Lemos, R. S., Strachan, R. A. and Topley, C. G., *Geol. Soc. London Spec. Publ.* **51**, 363–382.

O'Brien, S. J., Wardle, R. J. and King, A. F. (1983) The Avalon Zone: A Pan-African terrane in the Appalachian orogen of Canada. *Geol. J.* **18**, 195–222.

O'Brien, S. J., O'Neill, P. P., King, A. F. and Blackwood, R. F. (1988) Eastern margin of the Newfoundland Appalachians—a cross section of the Avalon and Gander zones. *Geol. Assoc. Can. Mineral. Assoc. Can. Field Trip Guidebook* **B4**, 1988 Annual Meeting, Memorial University of Newfoundland, St. John's.

Olszewski, W. J., Jr. and Gaudette, H. E. (1982) Age of the Brookville Gneiss and associated rocks, southeastern New Brunswick. *Can. J. Earth Sci.* **19**, 2158–2166.

Olszewski, W. J., Jr., Gaudette, H. E., Keppie, J. D. and Donohoe, H. V. (1981) Rb/Sr whole rock age of the Kellys Mountain Basement Complex, Cape Breton Island. *Geol. Soc. Am. Abstr. Programs* **13**, 169.

Poole, W. H. (1980) Rb-Sr age of some granitic rocks between Ludgate Lake and Negro Harbour, New Brunswick. In *Current Research, Part C, Geol. Surv. Canada Pap.* **80-1C**, 170–173.

Rast, N. (1979) Precambrian meta-diabases of southern New Brunswick—the opening of the Iapetus Ocean? *Tectonophysics* **59**, 127–137.

Rast, N. and Dickson, W. L. (1982) The Pocologan mylonite zone. In *Major Structural Zones and Faults of the Northern Appalachians*, eds. St-Julien, P. and Beland, J., *Geol. Assoc. Can. Spec. Pap.* **24**, 249–261.

Rast, N. and Skehan, J. W. (1983) The evolution of the Avalonian Plate. *Tectonophysics*, **100**, 257–286.

Ruitenberg, A. A. and Ludman, A. (1978) Stratigraphy and tectonic setting of early Palaeozoic sedimentary rocks of the Wirral-Big Lake area, southwestern New Brunswick and southeastern Maine. *Can. J. Earth Sci.* **15**, 22–32.

Ruitenberg, A. A., Giles, P. S., Venugopal, D. V., Buttimer, S. M., McCutcheon, S. R. and Chandra, J. (1979) Geology and mineral deposits, Caledonia area. *N. B. Dep. Natural Resour. Mem.* **1**.

Strong, D. F., O'Brien, S. J., Taylor, S. W., Strong, P. G. and Wilton, D. H. (1978) Aborted Proterozoic rifting in eastern Newfoundland. *Can. J. Earth Sci.* **15**, 117–131.

Stukas, V. (1977) *Plagioclase release patterns; a high resolution* $^{40}Ar/^{30}Ar$ *study*. Ph.D. thesis, Dalhousie University, Halifax, Nova Scotia.

Tanoli, S. K. and Pickerill, R. K. (1988) Lithostratigraphy of the Cambrian–early Ordovician Saint John Group, southern New Brunswick, *Can. J. Earth Sci.* **25**, 669–690.

Thomas, M. D. and Willis, C. (1989) Gravity modelling of the Saint George batholith and adjacent terranes within the Appalachian Orogen, southern New Brunswick. *Can. J. Earth Sci.* **26**, 561–576.

Wardle, R. J. (1978) *The stratigraphy and tectonics of the Greenhead Group: Its relation to Hadrynian and Palaeozoic rocks, southern New Brunswick*. Ph.D. thesis, University of New Brunswick, Frefericton.

Watters, S. E. (1987) Gold-bearing rocks—Bay of Fundy coastal zone. *N. B. Dep. Natural Resour. Inf. Circ.* **87-2**, 41–44.

Williams, H. (1979) Appalachian orogen in Canada. *Can. J. Earth Sci.* **16**, 792–807.

12 Palaeomagnetic and tectonic constraints on the development of Avalonian–Cadomian terranes in the North Atlantic region

G. K. TAYLOR and R. A. STRACHAN

12.1 Introduction

This chapter is an attempt to briefly synthesise the tectonic implications of the detailed geological summaries of individual areas presented in this volume and to integrate this with the available palaeomagnetic data to constrain a tectonic model for the evolution of the Avalonian–Cadomian terranes. We compare this model with previous models for the tectonic development of the North Atlantic terranes in late Precambrian–early Cambrian times.

Where options exist in interpreting the palaeomagnetic data we have elected to select reconstructed positions which emphasise the common origin of the Avalonian–Cadomian terranes and their affinity with Gondwanaland rather than a disparate origin. Furthermore we believe this reconstruction minimises the amount of plate movement required to attain relative plate positions in better established Cambrian and Ordovician reconstructions. While we fully recognise the existence of individual terranes within the Avalonian, southern Britain and Armorican composite terranes, the palaeomagnetic signatures of individual terranes have yet to be clarified and hence we can only treat them as composite blocks at present.

12.2 Palaeomagnetic constraints on a late Precambrian plate reconstruction

Our reconstruction for the Avalonian–Cadomian terranes in relation to Gondwanaland and Laurentia for the Period 600–580 Ma is shown in Figure 12.1 and discussed below.

12.2.1 *The main cratons*

The continuity of Africa and South America, after simple geometric closure of the Atlantic Ocean, has long been recognised (Wegener, 1912) and correlation of the Precambrian geology is well documented (for recent reviews see

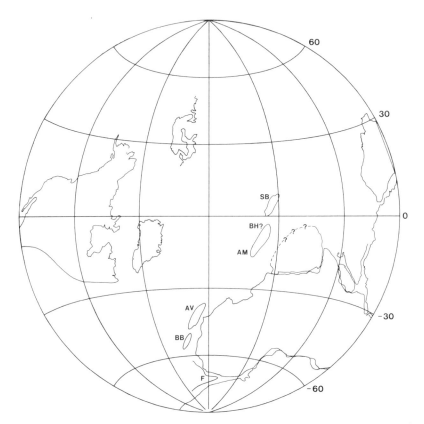

Figure 12.1 A reconstruction of the Avalonian–Cadomian terranes with respect to Laurentia, Gondwana and Fennoscandia based on palaeomagnetic data for the period 600–580 Ma (SB, Southern Britain; BH, Bohemia; AM, Armorica; AV, Avalon; BB, Boston Basin; F, Florida).

Torquato and Cordani, 1981; Rowley and Pindell, 1989; Dallmeyer, 1989a, b). Continuity of the Apparent Polar Wander (APW) paths for South America and Africa in this Pangea supercontinent configuration is also well established (e.g. Livermore *et al.*, 1985) and extends back until at least earliest Cambrian times (Perrin and Prevot, 1988). More controversial is the existence of a supercontinent comprising all the major continental plates throughout Precambrian times as argued for by Piper (1976, 1987) but dismissed by Van der Voo *et al.* (1984). Recent palaeomagnetic evidence (Kirschvink and Ripperdan, 1989) suggests that East (Antarctica, India, Australia and Asian fragments) and West (Africa, South America) Gondwanaland may have been separated at this time. This possibility however does not seriously affect the proposed reconstruction for the Pan-African terranes with respect to West Gondwanaland and North America and we have therefore adopted the most commonly proposed Gondwanaland reconstruction which juxtaposes East and West Gondwanaland.

Table 12.1   Reference poles for the Avalonian–Cadomian Terranes and Main Cratons

Symbol	Plate	Plat	Plong	Reference		Poles Used
NA	Laurentia	5	159	Watts et al.	(1980)	A, B, GD, CS, FD
AF	Gondwana	8	155	Perrin and Prevot	(1988)	31, 32, 35
FS	Fennoscandia	50	139	Piper	(1988)	4, 5, 6
AM	Armorica	30	250	Perigo et al.	(1983)	JC, KG1, SQ2, PP, Kh
SB	S. Britain	16	235	Piper	(1987)	6, 7, 8
AV	Avalonia	3	335	Irving and Strong	(1985)	MF, MG

Notes. Symbol is that used for reference in the text, Plat and Plong are the palaeomagnetic poles latitude and longitude respectively, and Poles Used refers to the numbers or letters used by the original authors to refer to individual poles and here meaned to produce the reference pole.

The selected reference pole for Gondwanaland (AF in Table 12.1) is derived from the compilation of Perrin and Prevot (1988) and is dated by them to earliest Cambrian times but includes a late Precambrian pole. Upper Proterozoic data from Gondwanaland is extremely scattered and does not follow a simple recognisable APW path. The two most likely causes for this are that there have been several discrete remagnetisation events associated with Proterozoic and Phanerozoic orogenic events which have overprinted the original primary remanence (Perrin and Prevot, 1988; Perrin et al., 1988) and that local block rotations may not be fully accounted for. Adoption of this reference pole results in an anticlockwise rotation of Africa and South America with the Mauritanides of Africa occupying mid to high palaeo-latitudes. Two recent results from the Eastern Sahara, from intrusions dated at 568 ± 13 and 589 ± 15 Ma (K-Ar), respectively, yield palaeolatitudes consistent with this reconstructed palaeoposition for Africa (Saradeth et al., 1989).

Florida is placed between South America and Africa based on the palaeontological, geochronological and palaeomagnetic evidence that suggests that Florida shares closer affinities with Gondwana than Laurentia in the early Palaeozoic (Opdyke et al., 1987). Furthermore this location is supported by the comparable Precambrian geological histories for these areas as discussed by Dallmeyer (1989b).

The data set for the North American craton during the late Precambrian–Cambrian has been collated by Watts et al. (1980), Dankers and Lapointe (1981) and Johnson and Van der Voo (1986). Although many studies have been carried out on rocks of this age in North America there remains considerable doubt as to the reliability of the ages assigned to magnetisations due to the possibility of widespread remagnetisation. The poles for the late Precambrian and early Cambrian however show little dispersion and cluster about a reference pole of 5°N 160°E (NA in Table 12.1). Johnson and Van der Voo (1986) and Van der Voo (1988) imply that North America would have consistently occupied near equatorial latitudes from about 600 Ma until 400 Ma, although its orientation varies during this period. The reconstructed position based on similar poles to reference pole NA was also used in previous reconstructions (Hagstrum *et al.*, 1980; Scotese, 1984; Wu *et al.*, 1986 and Piper, 1987). Contrary to this Bond *et al.* (1984, Figures 4b, c) place the North American craton in mid to high southerly palaeolatitudes at this time but note the disagreement with the palaeomagnetic data.

The Archaean to Phanerozoic palaeomagnetic database for Fennoscandia has recently been reviewed by Pesonen *et al.* (1989). They concluded that Fennoscandia acted as a single tectonic block during the late Precambrian–Cambrian period of interest, within the resolution of the palaeomagnetic data. From their interpreted APW paths they have constructed a drift history for Fennoscandia such that the area drifted northward from some 20°S at 850 Ma to a position 30°N by 600 Ma having undergone a near 90° anticlockwise rotation (Pesonen *et al.*, 1989, Figure 9). Piper (1988) has recently published new data and constructed a Late Proterozoic to early Palaeozoic APWP for Fennoscandia. This APWP is consistent with the basic model of northward drift proposed by Pesonen *et al.* (1989) and the reference pole FS is derived from these new results.

12.2.2   *The intervening terranes*

The APW path for Armorica, which is derived from localities in Brittany, Normandy and the Channel Islands, has remained little changed since reviewed by Perigo *et al.* (1983) although it has been supplemented by new Cambrian data and the inferred ages of magnetisations have been slightly amended by Taylor (1990). This path shows a continuity of poles from 600 to 500 Ma and infers a palaeolatitudinal change from a near equatorial position at 600 Ma to a 60–70°S position by 500 Ma for the Armorica microplate. Strachan *et al.* (1989) have recognised several discrete terranes within Armorica based on contrasting structural styles and chronological development and terrane boundaries marked by strike-slip faulting. The displacements between these terranes is however too small to be detected palaeomagnetically and the APW path is more consistent with Armorica acting as a single microplate throughout this period. Several authors have remarked upon the similarity of the Bohemian and Armorican APW paths and have suggested that Bohemia formed an integral part of the Armorican plate (Hagstrum *et al.*, 1980; Perigo *et al.*, 1983). Krs *et al.* (1987) have reported Middle and Upper Cambrian directions for Bohemia which place it at approximately 40°S at this time which would be consistent with Armorican data.

The late Precambrian–Cambrian exposures of England and Wales are confined to several small inliers whose relationship with each other is not fully established. At least three individual terranes have been recognised, namely the Monian of Anglesey (Gibbons, 1983) and the Charnian and Wrekin terranes of England and Wales (Pharoah *et al.*, 1987). The Monian terrane is itself a composite terrane consisting of at least three separate terranes (Gibbons, 1989a) and may extend into and include the Rosslare Complex of southern Ireland. Piper (1987) and Piper in Thorpe *et al.* (1984) has constructed an APW path which has two disjointed segments, the older dated at *c.* 700–650 Ma, the younger at 585–525 Ma. This path is interpreted as indicating that no significant translations of the separate blocks occurred and therefore that this area acted essentially as a single block during this period. However like the Armorican path it is not sufficiently detailed to exclude the possibility of rotation of individual blocks which might be expected given that these inliers are bounded by known strike-slip faults and may have significant displacements between terranes. The reference pole position (SB) is derived from poles from the Piper (1987) path dated at 585 Ma to earliest Cambrian, which includes the single available pole from Anglesey, and corresponds to this area occupying a near equatorial position at this time.

The palaeomagnetic data for the Avalon microplate (derived mainly from Newfoundland and Nova Scotia) have been collated by Johnson and Van der Voo (1986). They compared the observed palaeolatitudes for Avalon with stable cratonic North America for the period 620–400 Ma and concluded . that there is a consistent discrepancy between the two areas that suggests that Avalon lay 20–30° south of cratonic North America throughout this period. Furthermore they argue that the observed palaeolatitudes are consistent with reconstructions that place the Avalonian microplate at comparable palaeo-latitudes to the Armorican plate and northern Gondwana in Palaeozoic times. The reference pole AV used in this reconstruction was derived from the oldest poles available, the Marystown Group (Irving and Strong, 1985) dated as being between 625 and 600 Ma. Currently it is impossible to construct a coherent APW path for Avalon due to a wide scatter of the poles, and it is therefore not possible to determine if this is due to rotation of individual terranes/tectonic blocks within a larger composite single terrane or whether these terranes were more widely dispersed than is generally suggested by most authors based on geological evidence.

To the south of the Avalon zone lies the Boston basin which may form part of the main Avalon terrane or may itself be a separate terrane. Volcanics and clastic sediments of latest Precambrian to early Cambrian age have yielded a magnetisation which is believed to be primary and would indicate a palaeo-latitude of some 55° for this area at this time (Wu *et al.*, 1986). These authors suggest a southerly location close to the western margin of Gondwana and that this area forms part of an extended Avalonian microplate within the Avalonian–Cadomian belt.

12.2.3 *Comparison with previous reconstructions*

Several reconstructions of the Pan-African terranes, based on palaeomagnetic data, have been published in recent years (e.g. Hagstrum *et al.*, 1980; Bond *et al.*, 1984; Scotese, 1984; Wu *et al.*, 1986; Piper, 1987). These re-assemblies of the continents are marked by the diversity rather than the commonality of their solutions. This reflects the individual authors varying choices of reference pole positions for individual plates. This is in turn directly attributable to the inadequate data base for this time period (700–500 Ma), in terms of both the number and quality of available palaeomagnetic results. Furthermore the dipolar nature of the magnetic field means that existing pole positions are open to interpretation as to whether they are north or south poles and hence provide two possible locations for each land mass based purely on the choice of polarity. Although several new results have become available since the above mentioned reconstructions were made, these have done little to resolve the APW paths for the individual plates. Rather they have emphasised the need for caution in accepting palaeomagnetic results and in particular have highlighted the effects of later remagnetisation in these terranes, most of which have subsequently undergone one or more Phanerozoic orogenic events.

This reconstruction is essentially the same as that of Wu *et al.* (1986) and similar to that of Scotese (1984) with the exception that Baltica in this model was placed in a position between Laurentia and Gondwanaland. The major element shared by these models and those of Hagstrum *et al.* (1980) and Piper (1987) is the equatorial position of North America in late Precambrian or early Cambrian times. Bond *et al.* (1984) argue against the equatorial position for Laurentia, favouring a near polar position close to the margin of South America, primarily on the absence of widespread late Precambrian–Cambrian evaporite deposits in Laurentia and the affinity of Cambrian faunas in the Appalachians and northwest Argentina and a small number of Grenvillian (*c.* 900 Ma) radiometric ages from South America. While we are aware of these geological arguments we have opted for the more common restoration of Laurentia based on the palaeomagnetic data and the fact that this position would require minimal displacement of the area to achieve the relatively well constrained reconstructions of Cambrian and Ordovician times (Scotese, 1984; Livermore *et al.*, 1985).

12.3   Tectonic synthesis

In Figure 12.2 we present interpretative and simplified tectonostratigraphic columns for the Avalonian–Cadomian terranes discussed in this volume. The basement to these terranes, where exposed, is rather diverse in its nature. In West Africa and northwest France it is composed of basement gneisses of a continental character and approximately 2000–1800 Ma in age. The basement exposed in Nova Scotia, New Brunswick (?) and Bohemia is similarly continental in character, but of 'Grenvillian' age, *c.* 1200–900 Ma. Terranes

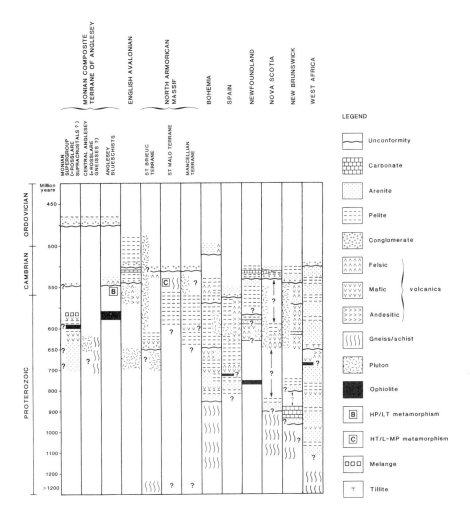

Figure 12.2   Interpretative and simplified Precambrian to early Palaeozoic tectonostratigraphic columns for the Avalonian–Cadomian terranes of the North Atlantic. (Compiled from chapters in this volume and with acknowledgements to Gibbons, 1989b.)

which do not contain exposed basement may be floored, in some cases, by a basement of oceanic or immature continental character (e.g. the Wrekin terrane of southern Britain, Pharoah *et al.*, 1987). Overlying Upper Proterozoic shallow marine platformal sequences are represented in New Brunswick, Nova Scotia and West Africa by quartzites and pelites (and carbonates in the case of New Brunswick). Evidence for platform collapse in New Brunswick is suggested by the carbonate breccias of the Martinon formation.

The latest Precambrian to early Palaeozoic record of the Avalonian–

Cadomian terranes, however, shows a number of common elements which are summarised as follows:

(a) the presence of thick volcano-sedimentary successions which consist mainly of turbidites in association with mafic–intermediate–acidic volcanics;

(b) widespread calc-alkaline plutonism, broadly coeval, and commonly co-genetic with the volcanics. Plutons have geochemical signatures suggestive of volcanic arc or marginal-continental settings;

(c) minor mafic and ultramafic plutonism, which may slightly pre-date but is generally overlapping in age with the calc-alkaline plutonism, possibly reflecting localised generation of oceanic-type crust (e.g. West Africa, Spain, Anglesey);

(d) evidence of major strike-slip displacements along steep, dominantly, sinistral shear zones (e.g. Anglesey, northwest France, New Brunswick, Nova Scotia);

(e) Cambro-Ordovician sedimentary sequences in all terranes are generally of fluviatile to shallow marine platform facies and contain Acado-Baltic faunas.

The timing of the onset of volcano-sedimentary deposition, calc-alkaline magmatism and deformation is broadly coeval but varies in detail from terrane to terrane and hence cannot be precisely correlated. The volcanic and plutonic sequences are consistent with those of modern arc-marginal basin assemblages within a subduction related setting. The age data and stratigraphic information (as shown in Figure 12.2) suggest a widespread record of volcanic arc magmatism, basin development and deformation between c. 700 and 550 Ma. We relate these events to southeastward-directed subduction beneath the Gondwanan craton. We envisage that the Avalonian–Cadomian terranes of the North Atlantic represent the dissected and largely allochthonous remnants of a complex mosaic of independently evolving volcanic arcs and marginal basins, some built on continental basement rifted from and lying outboard of the main Gondwanan craton represented in Africa. A broadly analogous setting might be that of the modern western Pacific margin where the Pacific plate is being subducted but creates a large number of individual arcs, trenches and marginal basins some of which are floored by continental basement while others have oceanic basements. The localised nature and irregular timing of events within the Avalonian–Cadomian terranes would be consistent with such a setting. Oblique subduction during at least part of the history of this subduction zone complex is implied by the strike-slip related nature of the deformation in most of the terranes.

The proposal that the evolution of the Avalonian–Cadomian terranes is related to a prolonged subduction event calls into question the usefulness of the term 'orogeny' as frequently applied to these terranes (i.e. Avalonian–Cadomian orogeny etc). The lack of regional deformation, large scale thrusting or crustal thickening suggest that the termination of this cycle of the tectonostratigraphic record of these terranes was not due to continent-

continent type collision (Murphy and Nance, 1989). This is further supported by the lack of regional angular unconformities between the subduction-related late Precambrian assemblages and the early Palaeozoic platformal sediments. Although this boundary is locally an angular unconformity (e.g. northwest France, southern Britain, Nova Scotia) in other areas it is a disconformity (e.g. New Brunswick, Newfoundland). The presence of an angular unconformity may simply reflect those parts of the belt which underwent localised deformation (Murphy and Nance, 1989). All the magmatic and tectonic events recorded within the Avalonian–Cadomian terranes can simply be viewed as the normal products of a complex subduction related arc-marginal basin system.

In common with previous models for the evolution of this belt (e.g. O'Brien *et al.*, 1983; Rast and Skehan, 1983; Nance, 1987) we envisage that separation of the Laurentia and Gondwana cratons commenced at around 800 Ma, resulting in rifting of the continental basement and the deposition and later subsidence of platformal sediments. We infer a period of *c.* 100 Ma during which opening of an ocean between the two cratons occurred. The tectonostratigraphic record within the Avalonian–Cadomian terranes implies that southeastward directed subduction beneath the Gondwanan craton commenced at *c.* 700–650 Ma, possibly diachronously. This tectonostratigraphic record of active subduction on the eastward margin of this ocean contrasts markedly with that preserved in Laurentia, where essentially passive margin sedimentation persisted throughout the Upper Proterozoic. The subduction of the eastern margin of this ocean is likely to have continued until *c.* 550–500 Ma. The observation that in some terranes regional strike-slip-related deformation is relatively late (*c.* 560–540 Ma, e.g. northwest France, Anglesey, and possibly also in New Brunswick and Nova Scotia) suggests that the termination of magmatic and tectonic activity in the Avalonian–Cadomian terranes may mark a change from orthogonal to oblique subduction and eventually to transcurrent motion along the margin of the Gondwanan craton.

Many problems remain to be resolved in making reconstructions and tectonic models such as these. For example two particular problems that we would choose to highlight are firstly the relationship between the ocean outlined in our model and the further evolution of the Iapetus ocean, and secondly on its potential linkage with the late Precambrian suture preserved in the Red Sea area. Such problems can only be answered by further detailed geological and palaeomagnetic studies of critical areas.

Acknowledgements

We would like to thank all the authors who have contributed to this volume and especially Wes Gibbons, Brendan Murphy and Damian Nance for preprints of papers in press. We are indebted to the numerous colleagues with whom we have had fruitful discussions about the Avalonian–Cadomian terranes. We thank Michelle D'Lemos for drafting the figures.

References

Bond, G. C., Nickeson, P. A. and Kominz, M. A. (1984) Breakup of a supercontinent between 625 Ma and 555 Ma: new evidence and implications for continental histories. *Earth Planet. Sci. Lett.* **70**, 325–345.

Dallmeyer, R. D. (1989a) Contrasting accreted terranes in the Southern Appalachian Orogen, basement beneath the Atlantic and Gulf Coast Plains, and western Africa. *Precambrian Res.* **42**. 387–409.

Dallmeyer, R. D. (1989b) The West African orogens and Circum-Atlantic correlatives. In *Avalonian and Cadonian Terranes of the North Atlantic*, eds. Strachan, R. A. and Taylor, G. K., Chapter 8, this volume.

Dankers, P. and Lapointe, P. (1981) Palaeomagnetism of Lower Cambrian volcanics and a cross-cutting Cambro-Ordovician diabase dyke from Buckingham (Quebec). *Can. J. Earth Sci.* **18**, 1174–1186.

Gibbons, W. (1983) Stratigraphy, subduction and strike-slip faulting in the Mona Complex of North Wales—a review. *Proc. Geol. Assoc. London* **94**, 147–163.

Gibbons, W. (1989a) Pre-Arenig terranes of northwest Wales. In *Avalonian and Cadonian Terranes of the North Atlantic*, eds. Strachan, R. A. and Taylor, G. K., Chapter 3, this volume.

Gibbons, W. (1989b) Britain and Ireland sutured—conference report. *Geology Today* **5**, 57–59.

Hagstrum, J. T., Van der Voo, R., Auvray, B. and Bonhomment, N. (1980) Eocambrian–Cambrian palaeomagnetism of the Armorican Massif, France. *Geophys. J. R. Astron. Soc.* **61**, 489–517.

Irving, E. and Strong, D. F. (1985) Palaeomagnetism of rocks from the Burin Peninsula, Newfoundland: hypothesis of the late Palaeozoic displacement of Acadia criticised. *J. Geophys. Res.* **90**, 1949–1963.

Johnson, R. J. E. and Van der Voo, R. (1986) Palaeomagnetism of the Late Precambrian Fourchu Group, Cape Breton Island, Nova Scotia. *Can. J. Earth Sci.* **23**, 1673–1685.

Kirschvink, J. L. and Ripperdan, R. L. (1989) A model of global plate motions for vendian and Cambrian time. *Geol. Assoc. Can. Mineral. Assoc. Can. Program Abstr.* **14**, A99.

Krs, M., Krsova, M., Pruner, P., Chvojka, R. and Havlicek, V. (1987) Palaeomagnetism, palaeogeography and the multicomponent analysis of Middle and Upper Cambrian rocks of the Barrandian in the Bohemian Massif. *Tectonophysics* **139**, 1–20.

Livermore, R. A., Smith, A. G. and Briden, J. C. (1985) Palaeomagnetic constraints on the distribution of continents in the late Silurian and early Devonian. *Philos. Trans. R. Soc. London* Ser. B **309**, 866–876.

Murphy, J. B. and Nance, R. D. (1989) Model for the evolution of the Avalonian-Cadomian belt. *Geology* **17**, 735–738.

Nance, R. D. (1987) Model for the Precambrian evolution of the Avalon terrane in southern New Brunswick, Canada. *Geology* **15**, 753–756.

O'Brien, S. J., Wardle, R. J. and King, A. F. (1983) The Avalon zone: A Pan-African terrane in the Appalachian orogen of Canada. *Geol. J.* **18**, 195–222.

Opdyke, N. D., Jones, D. S., MacFadden, B. J., Smith, D. L., Mueller, P. A. and Shuster, R. D. (1987) Florida as an exotic terrane: Palaeomagnetic and geochronologic investigation of lower Palaeozoic rocks from the subsurface of Florida. *Geology* **15**, 900–903.

Perigo, R., Van der Voo, R., Auvray, B. and Bonhommet, N. (1983) Palaeomagnetism of late Precambrian-Cambrian volcanics and instrusives from the Armorican Massif, France. *Geophys, J. R. Astron. Soc.* **75**, 235–260.

Perrin, M. and Prevot, M. (1988) Uncertainties about the Proterozoic and Palaeozoic polar wander path of the West African craton and Gondwana: evidence for successive remagnetization events. *Earth Planet. Sci. Lett.* **88**, 337–347.

Perrin, M., Elston, D. P. and Moussine-Pouchkine, A. (1988) Palaeomagnetism of Proterozoic and Cambrian strata, Ardar De Mauritanie, Cratonic West Africa. *J. Geophys. Res.* **93**, 2159–2178.

Pesonen, L. J., Torsvik, T. H., Elming, S. A. and Bylund, G. (1989) Crustal evolution of Fennoscandia—palaeomagnetic constraints. *Tectonophysics* **162**, 27–49.

Pharoah, T. C., Webb, P. C., Thorpe, R. S. and Beckinsale, R. D. (1987) Geochemical evidence for the tectonic setting of late Proterozoic volcanic suites in central England. In *Geochemistry and Mineralization of Proterozoic Volcanic Suites*, eds. Pharoah, T. C., Beckinsale, R. D. and Rickard, D., *Spec. Publ. Geol. Soc. London* **33**, 541–552.

Piper, J. D. A. (1976) Palaeomagnetic evidence for a Proterozoic supercontinent. *Philos. Trans. R. Soc. London Ser. A* **280**, 469–490.

Piper, J. D. A. (1987) *Palaeomagnetism and the Continental Crust*. Open University Press.

Piper, J. D. A. (1988) Palaeomagnetism of minor calc-alkaline intrusions, Fen Complex Southeast Norway. *Earth Planet. Sci. Lett.* **90**, 422–430.

Rast, N. and Skehan, J. W. (1983) The evolution of the Avalonian plate. *Tectonophysics* **100**, 257–286.

Rowley, D. B. and Pindell, J. L. (1989) End Palaeozoic–early Mesozoic western Pangean reconstruction and its implications for the distribution of Precambrian and Palaeozoic rocks around Meso-America. *Precambrian Res.* **42**, 411–444.

Saradeth, S., Soffel, H. C., Horn, P., Muller-Sohnius, D. and Schult, A. (1989) Upper Proterozoic and Phanerozoic pole positions and potassium-argon (K-Ar) ages from the East Sahara craton. *Geophys. J.* **97**, 209–221.

Scotese, C. R. (1984) An introduction to this volume: Palaeozoic palaeomagnetism and the assembly of Pangea. In *Plate Reconstructions from Palaeozoic Palaeomagnetism*, eds. Van der Voo, R., Scotese, C. R. *Geodyn. Ser.* **12**, 1–10.

Strachan, R. A., Treloar, P. J., Brown, M. and D'Lemos, R. S. (1989) Cadomian terrane tectonics and magmatism in the Armorican Massif. *J. Geol. Soc. London* **146**, 423–426.

Taylor, G. K. (1990) A palaeomagnetic study of two Precambrian–Cambrian dyke swarms from Armorica. In *The Cadomian Orogeny*, eds. D'Lemos, R. S., Strachan, R. A. and Topley, C. G., *Spec. Publ. Geol. Soc. London* **51**, 69–80.

Thorpe, R. S., Beckinsale, R. D., Patchett, P. J., Piper, J. D. A., Davies, G. R. and Evans, J. A. (1984) Crustal growth and late Precambrian–early Palaeozoic plate tectonic evolution of England and Wales. *J. Geol. Soc. London* **141**, 521–536.

Torquato, J. R. and Cordani, U. G. (1981) Brazil–Africa geological links. *Earth Sci. Rev.* **17**, 155–176.

Van der Voo, R. (1988) Palaeozoic palaeography of North America, Gondwana, and intervening displaced terranes: Comparisons of palaeomagnetism with palaeoclimatology and biogeographical patterns. *Bull. Geol. Soc. Am.* **100**, 311–324.

Van der Voo, R., Peinado, J. and Scotese, C. R. (1984) A palaeomagnetic re-evaluation of Pangea reconstructions. In *Plate Reconstructions from Palaeozoic Palaeomagnetism*, eds. Van der Voo, R., Scotese, C. R. and Bonhommet, N., AGU Washington D.C., *Geodyn. Ser.* **12**, 11–16.

Watts, D. R., Van der Voo, R. and Reeve, S. C. (1980) Cambrian palaeomagnetism of the Llano Uplift, Texas. *J. Geophys. Res.* **85**, 5316–5330.

Wegener, A. (1912) Die Entstehung Der Kontinente. *Geol. Rundsch.* **3**, 276–292.

Wu, F., Van der Voo, R. and Johnson, R. J. E. (1986) Eocambrian palaeomagnetism of the Boston Basin: Evidence for a displaced terrane. *Geophys. Res. Lett.* **13**, 1450–1453.

Index

Acado–Baltic fauna
  Avalon New Brunswick   214, 231
  Avalon Newfoundland   166, 168
  Avalon Nova Scotia   195, 208, 209
acritarchs
  Iberia   119, 120
andesite
  Armorica   85
  Avalon New Brunswick   226
  Avalon Newfoundland   173
  Avalon Nova Scotia   206
  Bohemian Massif   100, 101
  England and Wales   18
  Iberia   119
Anglesey   28–48
Appalachian–West African correlatives
  156–158
Armorica   65–92
  age of deformation   78–81
  Cadomian magmatism   74–76
  Cadomian terranes   65–92
  Icartian basement   69, 83
  Penthièvre complex   69–70
  phtanites, carbonaceous quartzites   76
  Post-Cadomian sedimentation   81–82
Avalon (zone) terrane, definition of in
  Newfoundland   166
Avalon of New Brunswick   214–236
  basement   220–221
  mylonite zones   228–230
  regional correlation   232–233
Avalon of Newfoundland   166–194
  bimodal volcanism   168, 174, 177
  Cambro-Ordovician stratigraphy   180–
  181
  Late Precambrian volcanics   172–174
  Latest Precambrian volcanics   177–178
  Mesozoic magmatism   182
  mid-Palaeozoic stratigraphy   181–182
  Palaeozoic magmatism   182
  plutonism   178–179
  Precambrian stratigraphy   171–178
  synthesis   186–189
  tectono-thermal history   182–186
Avalon of Nova Scotia   195–213
  basement   200–202
  constraints on the timing of deformation
  and metamorphism   201
  mid/late Proterozoic platformal rocks
  203–204

Avalon Superterrane in relation to
  Rosslare   62
backarcs
  England and Wales   22
  Iberia   118
basement
  Icartian of Armorica   69, 83
  Southern Appalachians   150–156
  West African orogens   134–136
Bassaride orogen   134–147
bimodal volcanism and/or dyke swarms
  Avalon New Brunswick   220, 227–228,
  230, 234
  Avalon Newfoundland   173, 174, 177
  Avalon Nova Scotia   205, 206, 208
  Iberia   117, 126
Blueschists
  NW Wales   28, 38–40
Bohemian Massif   93–108
  Brioverian unconformity with Cambro-
  Ordovician sediments   102
  Cadomian deformation   103
  Cadomian metamorphism and plutonism
  103
  correlation with the Brioverian of
  Armorica   95, 99–102
  overview   93–96
  terrane definition   95
Bohemicum terrane, Bohemia   95
Brioverian of the Bohemian Massif   99–
  102
Brioverian Supergroup, Armorica   72–74,
  76, 77, 85–86
Brunovistulicum terrane, Bohemia   95

calc-alkaline affinities
  Armorica   65, 67, 72, 74, 76–77, 85, 86,
  87
  Avalon New Brunswick   220, 225, 234,
  235
  Avalon Newfoundland   173, 174
  Avalon Nova Scotia   205, 206, 207
  Bohemia   99, 100, 102
  England and Wales   21
  Iberia   118, 119, 120, 126–127
  NW Wales   46
  Rosslare SE Ireland   55
  Southern Appalachians   158
  West Africa   140, 144, 146

Charnian Supergroup, Leicestershire 5,
15–18
Coedana Complex, NW Wales 36–38
cryptarchs
England and Wales 13

dacite
Avalon New Brunswick 226
Bohemian Massif 100, 101
Iberia 119
diorite
Armorica 69, 74, 75, 86
Avalon New Brunswick 231
Avalon Newfoundland 178
Avalon Nova Scotia 200, 205, 207
England and Wales 15, 17
NW Wales 40
dubiofossils 17, 19
dyke swarms
Armorica 72, 84
Avalon New Brunswick 220, 222, 227–228
Avalon Newfoundland 176
Avalon Nova Scotia 206
Iberia 113
Rosslare SE Ireland 56, 61

eclogite, Iberia 125
Ediacaran fauna
Avalon Newfoundland 176
England and Wales 16, 18, 19

foliated plutonic rocks
Armorica 84
Iberia 127
NW Wales 40
forearc basins and environment
England and Wales 21
NW Wales 45
fossil assemblages
Cambrian faunas, New Brunswick 229,
231
Cambrian faunas, Newfoundland 172,
174, 181
Cambro-Ordovician fauna, Nova Scotia
198, 208
early Cambrian fauna, West Africa 140
Riphean faunas, Iberia 115, 120, 122
trilobite fauna, of Southern Appalachians
148
Vendian fauna, Iberia 122
see also Acado-Baltic fauna, Rhenish-Bohemian fauna, Laurentian fauna,
dubiofossils, acritarchs, cryptarchs,
microfossils, stromatolites

gabbro
Armorica 67

Avalon Newfoundland 171, 178, 179
Bohemia 100
Rosslare SE Ireland 55
glaciogenic deposits
Avalon New Brunswick 234
Avalon Newfoundland 175
West Africa 140
granites and granitoids
Armorica 67, 74–75, 76, 77, 85
Avalon New Brunswick 214, 224–225,
230
Avalon Newfoundland 170, 171, 178,
179, 184
Avalon Nova Scotia 200, 205
Bohemia 105
Iberia 113, 115, 120
NW Wales 36, 40
Southern Appalachians 150
West Africa 140
granodiorite
Armorica 69, 77
Avalon Newfoundland 178
NW Wales 40
granulites
Bohemia 97
Iberia 125

Iberia Massif 109–133
correlation with Cadomian 113
Iberian terrane, tectono-thermal
evolution 130
Precambrian sequences and suspect
terranes 111–113
stratigraphy of Precambrian sequences in
the Iberian terrane 113–122
introduction 1–3
island arc tholeiites
England and Wales 21

Laurentian faunal province 209
limestone and carbonates
Armorica 86
Avalon New Brunswick 221, 222
Avalon Newfoundland 168, 180
Bohemia 97
England and Wales 14
Iberia 122
NW Wales 35
Llangynog Inlier, Dyfed 18
Llŷn Peninsula 28–48
Longmyndian Supergroup 5–27
sedimentology 11–12
stratigraphy 7–8
structure of 9–10

magmatic arcs
Avalon Nova Scotia 195, 198, 210, 211
England and Wales 14, 17, 22, 23

Malvernian plutonic complex 5
Mauritanide orogen 134 – 147
melanges
  NW Wales 28, 35, 43
microfossils
  Avalon Newfoundland 176
  Bohemian Massif 96, 102
  England and Wales 13
  West Africa 140
migmatites
  Armorica 75, 76
  Bohemian Massif 97, 98
Moldanibicum terrane, Bohemia 95
Mona Supercomplex 28 – 48
  age of 31, 43
Monian Supergroup 33 – 36
mylonite
  Avalon Newfoundland 168
  NW Wales 31

oceanic tholeiite, Avalon Newfoundland 171
Old Radnor Inlier, Welsh Borderland 12 – 14
ophiolites and oceanic crust
  Avalon Newfoundland 172
  Bohemian 100
  England and Wales 22
  Iberia 110, 111, 115, 128, 130
  metabasites 38
  NW Wales 31, 45
  West Africa 146
Osceola granite, Southern Appalachians 150 – 152

palaeomagnetic and tectonic constraints on the tectonic development of Avalonia-Cadomia 237 – 247
Pan-African orogenic terranes 135
Pangea 1, 159, 238
plate reconstruction 1, 134, 210, 237 – 242
pre-Arenig terranes of NW Wales 28 – 48
Precambrian of England and Wales 5 – 27
  depositional setting 21 – 24
  sediments 19 – 21, 5 – 27
Precambrian terranes in Iberia 109 – 133
pyroclastics
  Armorica 75
  Avalon New Brunswick 226, 230
  Avalon Newfoundland 174
  Avalon Nova Scotia 205
  England and Wales 9, 10, 16, 17
  Iberia 119

red beds
  Armorica 65, 69, 81 – 82
  Avalon New Brunswick 232, 235
  Avalon Newfoundland 176 – 177, 183
  England and Wales 21
  Iberia 120
  West Africa 140, 141
regional correlation
  Armorica 83 – 86
  NW Wales 40 – 41
  see also Chapters 9 – 12
Rhenish – Bohemian fauna
  New Brunswick 215
  Nova Scotia 209
Rhenohercynicum terrane, Bohemia 95
rhyolite
  Armorica 74 – 75
  Avalon Newfoundland 173, 174, 178
  Avalon Nova Scotia 206
  Bohemian Massif 100
  England and Wales 18
  Iberia 119
Rokelide orogen 134 – 147
Rosslare Complex, SE Ireland 49 – 64
  age 51
  Caledonian intrusions 61, 62
  definition and history 51 – 57
  marginal rocks 57 – 61
  mylonite zones 49, 55
  relation to Icartian of Cadomia 51
  relation to Monian, NW Wales 51

Sarn Complex, NW Wales 40
Saxothuringicum-Lugicum terrane, Bohemia 95
serpentinite
  Avalon New Brunswick 224
  Avalon Newfoundland 171
  Bohemian Massif 98, 100
  Iberia 115, 128
  NW Wales 28, 34
  Southern Appalachians 150
  West Africa 137
Southern Appalachian orogen 146 – 160
  basement of 150 – 156
  Osceola granite 150 – 152
  terrane accretion 158 – 160
  trilobite fauna 148
Stanner-Hanter Complex 5, 12, 14
strike slip-fault systems and shear zones
  Armorica 65, 67, 76, 78 – 80, 87
  Avalon New Brunswick 214
  Avalon Newfoundland 171
  Avalon Nova Scotia 204
  Bohemian Massif 103
  England and Wales 5, 6, 8, 9, 12, 14, 23
  Kilmore-Wilkeen mylonite zone, Rosslare 51
  NW Wales 28, 31, 32, 37, 42
stromatolites
  Avalon New Brunswick 221, 232
  Avalon Newfoundland 171

stromatolites—*continued*
　Avalon Nova Scotia　203
subduction
　Armorica　65, 77
　Avalon New Brunswick　226, 230
　England and Wales　21, 22, 23, 28
　Iberia　120, 128, 130

tectonic setting
　Armorica　86–88
　Avalon New Brunswick　232–234
　Avalon Newfoundland　186–189
　Avalon Nova Scotia　210–211
　Iberia　127–130
　Monian of NW Wales　44–46
　Precambrian sediments of England and
　Wales　21–24
　Rosslare Complex　62
　West Africa　145–147
tectonic synthesis of the Avalonian–
　Cadomian belt　242–245
terranes, definitions and tectonics
　Armorica　65–67
　Avalon Composite terrane in Nova
　Scotia　195–213
　Avalon Zone in Newfoundland　166
　Avalon Zone in Nova Scotia　195–199
　Bohemian Massif　95
　Iberia　110–111
　individual terranes and their tectonics
　accretion of, in the southern
　Appalachians　158–160
　blueschist terrane of Anglesey NW
　Wales　38–40, 44
　Carolina terrane of Southern
　Appalachians　148
　Coedana Complex of NW Wales　36–
　38, 44
　Iberian　110, 113–122
　Mancellian terrane, North Armorica
　77
　Monian Supergroup of NW Wales
　33–36, 44
　Pan-African　135
　Piedmont terranes of Southern
　Appalachians　148–150
　Sarn Complex, NW Wales　40, 44
　South Portuguese　110
　St. Brieuc terrane, Armorica　67–75
　St. Malo terrane, Armorica　76–77
　NW Wales　31

terranes, palaeomagnetic and tectonic
　constraints on their evolution　238–248
terrestrial clastic deposits
　Avalon Newfoundland　168, 174, 176–
　177, 184
　see also red beds
tholeiitic affinities
　Armorica　72, 74
　Avalon New Brunswick　226, 229, 230
　Avalon Newfoundland　168, 173
　Avalon Nova Scotia　206, 207
　Bohemia　99, 100
　Iberia　117, 118
　NW Wales　35
　Rosslare, SE Ireland　57
　West Africa　137, 139, 141, 146
tonalite
　Armorica　75
　Avalon Nova Scotia　205
　Iberia　120
　NW Wales　40
trace fossils, New Brunswick, Saint John
　Group　231
turbidites
　Armorica　67, 73, 74, 77, 85, 86
　Avalon Newfoundland　168, 174, 175
　Avalon Nova Scotia　195, 206
　Bohemia　101
　England and Wales　10, 11, 12, 16, 22
　Iberia　118, 119, 122
　NW Wales　33

Uriconian Volcanic Complex　5, 8, 13, 14

Valencia　de las Torres-Cerro Muriano
　Supergroup　117–119
volcanic arc
　Armorica　72
　Avalon New Brunswick　233
　Avalon Newfoundland　173
　Avalon Nova Scotia　205, 206
　England and Wales　21
　Iberia　127

West Africa　134–165
　basement　136
　geological setting　136–141
　tectono-thermal history　142–144
　terranes　134–165
West African-Appalachian correlatives
　156–158